澜湄职业教育培训中心暨柬埔寨鲁班工坊系列教材
A Series of Textbooks for Lancang-Mekong Vocational Education Training Center and Cambodia Luban Workshop

车削基础训练（高级）

Basic Training of Turning (Advanced)

主　编　方　力

Editor in chief　　Fang Li

副主编　王　慧

Deputy editor in chief　　Wang Hui

西安电子科技大学出版社

Brief Introduction

This book is compiled according to the requirements of technical skills and theoretical knowledge for the advanced lathe worker in the revised national vocational qualification standard. This book introduces four modules, including machining long lead screw and worm, eccentric parts and multi-throw crankshaft, machining gearbox, and machining assembly, with tasks and subtasks in each module, with regard to the workpiece diagram recognition, processing analysis and calculation of various parts, such as long lead screw, multi-start thread, multi-line worms, double eccentric parts, four-throw crankshaft, gear reducer, worm gear reducer, bevel gears, bisector parts, mold, combined shaft and kit, as well as their clamping and positioning methods, processing methods, dimensional and geometric accuracy test methods, surface quality assurance methods, common quality problems and preventive measures or solutions, etc.

This book can be used as a teaching book for mechanical, mechatronics,etc. application-oriented undergraduates in higher vocational colleges, as well as a reference book or skill training book for relevant technicians.

图书在版编目（CIP）数据

车削基础训练 ： 高级 ＝ Basic Training of Turning (Advanced) / 方力主编. -- 西安：西安电子科技大学出版社，2025.4. ISBN 978-7-5606-7607-4

Ⅰ. TG51

中国国家版本馆 CIP 数据核字第 2025ZB7525 号

书　　名　车削基础训练(高级)
　　　　　　CHEXIAO JICHU XUNLIAN(GAOJI)

策　　划　秦志峰
责任编辑　郭　静
出版发行　西安电子科技大学出版社（西安市太白南路 2 号）
电　　话　(029) 88202421　88201467　邮　编　710071
网　　址　www.xduph.com　　　　　电子邮箱　xdupfxb001@163.com
经　　销　新华书店
印刷单位　西安日报社印务中心
版　　次　2025 年 4 月第 1 版　　　2025 年 4 月第 1 次印刷
开　　本　787 毫米×1092 毫米　1/16　印　张　11.5
字　　数　195 千字
定　　价　32.00 元
ISBN 978-7-5606-7607-4
XDUP 7908001-1
＊＊＊ 如有印装问题可调换 ＊＊＊

General Foreword

Serving the Belt and Road Initiative of China, the Lancang-Mekong Vocational Education Training Center and Cambodia Luban Workshop is a joint project undertaken by Tianjin Sino-German University of Applied Sciences(TSGUAS) for the Ministry of Foreign Affairs, the Ministry of Education and the Tianjin Municipal People's Government. Based in Cambodia, the project is designed to serve five countries in the Lancang-Mekong area and radiate to other ten ASEAN countries. It integrates functions of vocational training, vocational education, scientific research, cultural inheritance and innovation&entrepreneurship, develops both academic and non-academic education, and operates as a market-oriented international vocational training center.

At the initial stage of the project, 18 training rooms including mechanical processing technology, electrical technology and communication technology were built in three training centers for mechatronics and communication technology majors, with a total construction area of 6,814 square meters and more than 1,600 sets of equipment.

The project will implement a "three-phase" plan. Based on the specialty construction in the first phase, international tourism, logistics engineering, automobile maintenance, building electricity and other specialties will be set up in the second phase to carry out technical skills training for Chinese & Cambodian enterprises and Cambodian people. Meanwhile, higher vocational education, applied technology undergraduate education, joint postgraduate education and other academic educations will be carried out to explore systematic talents cultivation of "medium and high vocational education, undergraduate education, and postgraduate education for a master's and doctoral degree".

Since 2017, as many as 95 articles about the project have been published by mainstream media including People's Daily, Guangming Daily, China Education News, Xinhuanet, etc. from home and abroad. After over two months of field study and research, Tianjin Television produced two feature stories named "Khmer Training", each lasting 30 minutes. The two episodes were broadcast on May 6th and May 13th, 2019 respectively, featuring "on and on sails the vocational education, overseas shines the Luban Workshop". They give a full coverage of how TSGUAS teachers brought advanced skills to local areas and how friendship flourished along the Belt and Road Initiative route—a great contribution to the BRI. On July 18th, 2019, the Royal Government of Cambodia conferred the Officer of the SAHAMETREI Medal to the secretary of the Party Committee of TSGUAS, and the Knight of the SAHAMETREI Medal to the President and Vice President in charge of this project, with the signature of Prime Minister Hun Sen of Cambodia. On July 22, 2019, China Education Association for International Exchange awarded TSGUAS the medal of "Featured Cooperation Project of China-ASEAN Higher Vocational Colleges". In October, 2019, the President of National Polytechnic Institute of Cambodia (NPIC) presented 11 teachers with certificates and medals for their outstanding contributions to the Ministry of Labor and Vocational Training of Cambodia. Tianjin Sino-German University of Applied Sciences together with National Polytechnic Institute of Cambodia (NPIC) and their partners with enterprises was approved as the Belt and Road Joint Laboratory (Research Center)—Tianjin Sino-German and Cambodia Intelligent Motion Device and Communication Technology Promotion Center in December, 2020.

The Center has become a training base in Langcang-Mekong areas for technical talents training, a talent support base for Chinese enterprises overseas, a demonstration base for international students, and a base for teachers training. The Center is a key educational project of the Ministry of Foreign Affairs to serve the Belt and Road Initiative with foreign participation and entity institutions involved locally. The project will serve the social-economic and cultural development of the countries along the Initiative, enhancing the well-being of mankind; it will also serve the production output capacity of Chinese enterprises to help national development as well as enhance the international development of vocational education and the quality of its connotation. The project is a bridge connecting vocational education of Tianjin with the world, which marks a new stage of the city's international exchange and cooperation from a lower-medium to a medium-higher level.

The team of the project has compiled a series of textbooks for training, involving six occupations (electrotechnics, lathe, milling, CNC operation, bench and 4G communication network) from elementary, intermediate to advanced level based on current human resources situation in Langcang-Mekong countries, China's teaching equipment, and Chinese vocational qualification standards. These 19 textbooks target competence development and orient students to work tasks, combining theory with practice, and learning with practicing so as to put knowledge and skills into real situations. The textbooks aim to provide skills standards for the six occupations and lay foundations for the upgrading of the technological level of Lancang-Mekong countries.

ZHANG Xinghui

Party Secretary of Tianjin Sino-German University of Applied Sciences

March, 2021

A Series of Textbooks for Lancang-Mekong Vocational Education Training Center and Cambodia Luban Workshop Editorial Committee

Preface

This book aims at providing high-quality technical skills training and educational concepts to Lancang-Mekong countries so as to better serve the Belt and Road Initiative and the Lancang-Mekong Vocational Education and Training Center. Adhering to the modern apprenticeship talent training model and with the guidance of requirements of technical skills and theoretical knowledge for the advanced lathe workers in the newly revised national vocational qualification standard, this book is compiled with reference to the actual production of the enterprise and the actual operation of skilled workers, combined with the development level of the local manufacturing industry and the training equipment used.

In terms of contents, this book strives to highlight practicability and operability, ability-oriented and guided by practical work tasks, taking projects as the carrier. It adheres to the principle of project-based teaching mode and combines theory with practice, focusing on practice, supplemented by theory with sufficient introduction to theoretical knowledge. With the emphasis on practice and the cultivation of hands-on ability, the book is compiled through practicing according to the requirements of practical tasks, from shallow to deep, from easy to difficult. The scientific education theory of systematic working process is flexibly applied, and the knowledge and skills that students need to master are fully reflected in each module and specific work tasks.

The book designs four modules: machining long lead screw and worm, machining eccentric parts and multi-throw crankshaft, machining gearbox and machining assembly, with tasks and subtasks in each module. At the beginning of each task, the theoretical basic knowledge required in the task is compiled, so as to combine theory with practical work, and to make the book suitable for readers with or without professional foundation.

The modules and tasks designed in this book are all from practical production cases and skill appraisal, and can meet the requirements and regulations of theoretical knowledge and related skills for the advanced lather in national vocational qualification standard. In order to in line with international teaching materials, this book is written in Chinese and English. It can not only meet the learning needs of domestic readers for English of machinery related majors, but also be applied to the training of overseas related skills and the self-study of foreign readers.

This book is written by FANG Li and WANG Hui, teachers from Tianjin Sino German University of Applied Sciences. Among them, Module 1, 2 and 3 are written by WANG Hui, Module 4 is written by FANG Li. This book's follow-up work is finished by XIANG Dan, a teacher of Tianjin Sino German University of Applied Sciences. Due to our limits, welcome to put forward your suggestions and advices.

Writers

September, 2024

Contents

Module One　Machining Long Lead Screw and Worm 1

　Task 1　Machining Long Lead Screw 1

　　Subtask 1　Lead Screw Processing and Test 1

　　Subtask 2　Turning Long Lead Screw 11

　Task 2　Machining Multi-start Thread and Multi-line Worms 21

　　Subtask 1　Machining Multi-start Thread 21

　　Subtask 2　Machining Multi-line Worms 31

Module Two　Machining Eccentric Parts and Multi-throw Crankshaft 43

　Task 1　Machining Double Eccentric Parts 43

　　Subtask 1　Machining Double Eccentric Sleeve 43

　　Subtask 2　Turning Coaxial Double Eccentric Shaft and

　　　　　　　　Reverse Double Eccentric Sleeve 52

　Task 2　Machining Four-throw Crankshaft 67

Module Three　Machining Gearbox 90

　Task 1　Machining Gearbox Body 90

　　Subtask 1　Machining Gearbox Hole 90

　　Subtask 2　Measuring Gearbox Size 96

　Task 2　Machining Worm Gear Housing 106

　Task 3　Machining Bevel Gears 113

Module Four　Machining Assembly 122

　Task 1　Machining Bisector Parts 122

　　Subtask 1　Machining Bearing Bush 122

　　Subtask 2　Machining Split Bearing Seat 127

Task 2　Machining Mold 134

　　Subtask 1　Machining Cylinder Mold 134

　　Subtask 2　Machining Gear Mold.. 143

Task 3　Machining Shaft and Kit Combination.. 147

　　Subtask 1　Machining Tri-eccentric Shaft Sleeve.................................. 147

　　Subtask 2　Machining Cone Eccentric Four-item Assembly 159

References 174

Module One   Machining Long Lead Screw and Worm

🌀 Task 1   Machining Long Lead Screw

Subtask 1   Lead Screw Processing and Test

> 【Knowledge and Skills Objectives】
>
> (1) Master knowledge of turning long lead screw.
>
> (2) Master the selection of geometric parameters for precision turning tools for trapezoidal thread (screw).
>
> (3) Master the precision test method for long lead screw.

【Related Knowledge】

I . Workpiece diagram recognition of trapezoidal screw

The diagram of trapezoidal screw is shown as in Figure 1.1. The left end of the screw is a short shaft with a diameter of $\phi 18_{-0.06}^{-0.018}$ mm and with a length of 25 mm, and the right end is a trapezoidal thread with Tr28 × 5 mm.

II . Brief introduction to long lead screw

1. Category

There are three commonly-used screws: sliding screw, rolling screw and static screw. The sliding screw is the most widely used and its processing is convenient. The friction of the rolling screw is small, the transmission accuracy of it is high, but the processing of it is complex. The static screw has high transmission accuracy and high bearing capacity, but its manufacturing process is complex, which is mostly used for the transmission of precision screw with large diameter.

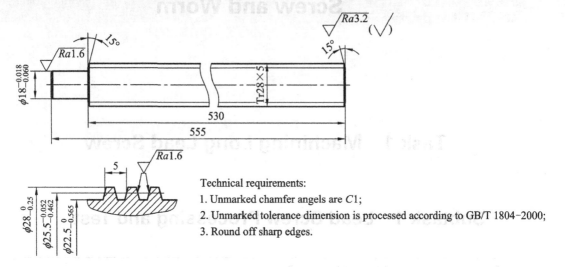

Figure 1.1   Diagram of trapezoidal screw

Technical requirements:
1. Unmarked chamfer angels are C1;
2. Unmarked tolerance dimension is processed according to GB/T 1804-2000;
3. Round off sharp edges.

2. Processing requirements of long lead screw

(1) The accuracy of machine tools should be enhanced to reduce machining errors. Usually, machine tools with high accuracy and small wear are selected, and the positioning benchmarks of machine tools should be adjusted to improve the accuracy of machine tools.

(2) The tool should be selected correctly according to the material of the workpiece, and the influence and the strength of the tool on turning should be considered when the tool forms the thread angle.

(3) Reasonably choose turning methods and cutting parameters. The workpiece ought to be fully cooled and lubricated to reduce the impact of workpiece deformation.

III. Precautions for processing long lead screw

The diagram of lathe lead screw (a kind of sliding screw)(see Figure 1.2) is a kind of flexible workpiece with a large length diameter ratio, which requires high precision. The lead screw is easy to deform under the action of external and internal stress. Therefore, attention should be paid to the deformation during machining.

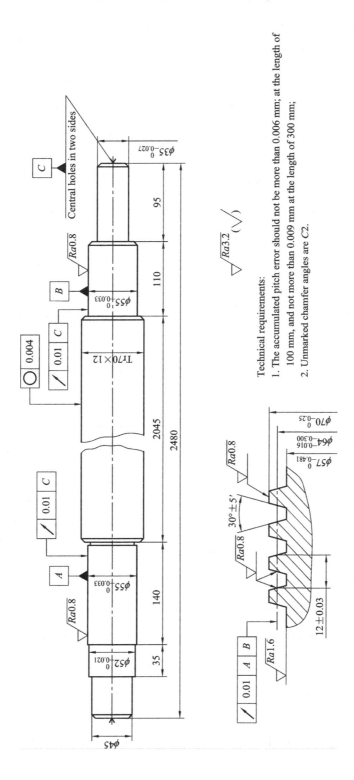

Technical requirements:

1. The accumulated pitch error should not be more than 0.006 mm; at the length of 100 mm, and not more than 0.009 mm at the length of 300 mm;

2. Unmarked chamfer angles are C2.

Figure 1.2   The diagram of lathe lead screw

1. The screw material should be guaranteed to meet requirements

The screw material should have enough strength and stable structure, good wear resistance, and appropriate hardness and toughness to ensure that the cutting process does not affect the machining accuracy and surface roughness. For unhardened screw, the quenched and tempered 45 steel is the common material for ordinary screw; high quality carbon tool steel (such as T10A, T12A, etc.) is the common material for precision screw. For the screw that needs to be hardened, the material is usually alloy steel, such as 9Mn2V, GCr15, etc.

2. The precision of positioning benchmarks should be guaranteed

It is necessary to ensure the precision of the center hole and the outer circle in the machining process. The center hole needs to be grinded (or milled) so that it has high shape accuracy and sufficient contact area, and it can cooperate well with the rotary tip of the machine tool. In the process of screw machining, the center hole should be continuously repaired and grinded.

3. Eliminate the internal stress which serves as the main factor causing screw deformation

In order to prevent and reduce screw deformation, ball processing should be strictly carried out to achieve a stable structure, and the aging treatment should be arranged reasonably in the machining process to eliminate the internal stress generated in the cutting process. In addition, during the heat treatment and mechanical processing flow between each process, the screw should be suspended vertically to avoid bending deformation due to the weight of the screw itself.

4. The screw clamping method should be improved

The clamping method of screw is shown in Table 1.1.

5. The accumulative pitch error of thread finish turning is analyzed

The accumulative pitch error directly affects the accuracy of the precision screw transmission movement, which is a key issue that should be paid attention to in processing. There are generally three factors affecting the accumulative pitch error:

(1) The temperature difference between the workpiece and the machine tool screw.

Table 1.1   The clamping method of screw machining

No.	The clamping method	The analysis for reducing deformation
1	The tail is clamped with a movable center(the lead screw can be freely extended after being heated)	
2	Clamped with the spring(the lead screw can be freely extended to the tail after being heated)	
3	There is no center at the tail(the lead screw can be freely extended to the tail after being heated)	

Note: After the lead screw is heated and elongated, the accumulative pitch error will occur, which can be solved by compensation.

(2) The bed guide rails of the machine tool are not parallel in the horizontal plane, which makes the machined thread have conicity, resulting in the diameter error and the accumulative pitch error. The calculation formula for the maximum accumulative pitch error S is

$$S = \frac{\Delta d_{max}}{2} \times \frac{\tan \alpha}{2} \tag{1.1}$$

where, Δd_{max} represents the maximum diameter difference， unit in mm. α represents the thread profile angle, unit in (°).

(3) The machine bed is twisted, and the guide rail is tilted in the vertical plane.

6. Measures should be taken to reduce the pitch error

In order to reduce the accumulative pitch error, the finish turning should be carried out in a constant temperature room, so as to keep the temperature difference between the thread and the lead screw less than ΔT (that is, the temperature difference between the cutting area and the surrounding environment). The thread length of the precision screw in Figure 1.2 is 2480 mm, the thread accuracy is level 6, and the accumulative pitch error tolerance over the total length is 0.02 mm.

Therefore, the temperature difference between the thread and the screw ΔT is

$$\Delta T = \frac{\Delta L}{aL} \tag{1.2}$$

where, ΔL represents the pitch tolerance; a represents the linear expansion coefficient of steel, $°C^{-1}$; L represents the part length.

7. The screw should be grinded.

Grinding is a method to improve the pitch accuracy of precision lead screw. Grinding has a good effect on reducing single pitch error, periodic error and short-distance accumulative pitch error of the screw.

During thread grinding, a grinding agent is injected between the grinding tool and the screw spiral surface. Due to the relative spiral motion, the abrasive particles in the grinding agent are subjected to frictional and compressive forces. Therefore, part of the abrasive particles will be embedded in the surface of the screw. Under the action of friction and extrusion pressure, it can produce small cutting effect which can reduce the various errors on the screw surface.

IV. Selection of geometric parameters and cutting parameters of fine turning tool for long lead trapezoidal thread

1. Selection of geometric parameters of fine turning tool

The geometric parameters of fine turning tool for long lead trapezoidal thread are shown in Figure 1.3. The requirements of fine turning tool are as follows.

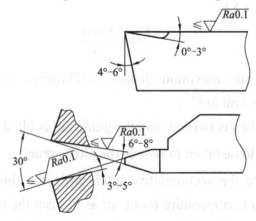

Figure 1.3   The geometric parameters of fine turning tool

(1) High hardness and good wear resistance are required. Due to the long continuous cutting length of the tool when turning the lead screw, the low wear resistance of the tool will inevitably affect the thread accuracy, so fine-grained YG6

or YG6X hard alloy is used as the material for fine turning tool.

(2) The cutting edge of the tool ought to be sharp. When fine-turning thread, the amount of back engagement is very small, and the cutting load is concentrated near the cutting edge. If the cutting edge of the tool is not sharp, normal cutting cannot be performed, which will not only aggravate the wear of the tool, but also increase the cutting heat, deform the workpiece and reduce the machining accuracy.

(3) The surface roughness value of the tool ought to be small. If the tool surface roughness value is large, on one hand, the uneven cutting edge directly affects the surface roughness of the machined surface; on the other hand, the tool itself is easy to wear. Therefore, the surface roughness values of the front and back of the fine turning tool should generally be no larger than $Ra0.8$ μm.

2. Selection of cutting parameters of fine turning tool

When processing the lead screw, the cutting parameters of each process must be strictly controlled. If the cutting speed is too high or the amount of back engagement is too large, the load on the tool will be increased, the tool wear will be accelerated, the accumulative error of the pitch will be increased, and the cutting heat will also be increased. For semi-finish turning of the lead screw, the cutting speed should be 2-3.5 m/min, and the cutting back engagement should be 0.05 mm; for finish turning of the lead screw, the cutting speed should be 1-1.5 m/min, and the cutting back engagement should be 0.02-0.01 mm.

V. Precision test of long lead screw

The requirements of precision test of long lead screw are as follows:

(1) The long lead screw must not be processed by cold turning (internal stress will be generated during the forging process of the workpiece, and the original state will still be restored if the stress is not removed by heat treatment after turning).

(2) The radial runout of all outer circles is not more than 0.04 mm.

(3) If the precision level of the lead screw is different, the inspection method is different. The precision of the screw is divided into six levels, namely 4, 5, 6, 7, 8, and 9, and the precision is reduced in order. The screw precision indexes are the spiral error and the pitch error. In production and processing, for low-precision screw, its pitch is tested by a special template; for medium-level precision screw, its

pitch is tested by a screw pitch measuring instrument; for high-precision screws, its pitch is tested by a JCO-30 screw inspector.

The principle of the JCO-30 screw inspector is shown in Figure 1.4. The instrument consists of a reading microscope 2, a precision scribing ruler 3, rail 1, etc. The screw inspector needs to work in conjunction with a workbench, a measuring meter, etc. When measuring, the lead screw is clamped between the two centers, and the precision scribing ruler is connected in series with the workpiece, and is in contact with the screw profile on one side. When measuring, the microscope is first aligned to the zero position of the precision scribing ruler 3, and the pointer of the micrometer dial gauge 6 is adjusted to zero. Then, the rotation probe 5 exits from the screw tooth surface, and then moves the workbench, so that the precision scribing ruler 3 is precisely aligned with the precise line of the corresponding pitch deviation(nP). At this time, the probe enters the required tooth surface and the pitch deviation, ΔnP, can be read from the micrometer dial gauge.

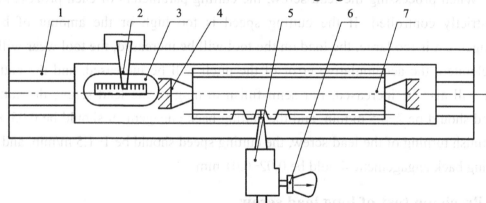

1-rail; 2-reading microscope; 3-precision scribing ruler; 4-top; 5-rotation probe;
6-micrometer dial gauge; 7-lead screw.

Figure 1.4   The JCO-30 screw inspector

The screw inspector is a detection instrument, which can be used to measure the screw pitch error, the maximum value of adjacent pitch error, and the maximum accumulative pitch error of the screw within a certain length and the full length.

【Skills Training】

Turning and Inspect Lead Screw

Processing the trapezoidal thread lead screw (written as lead screw for short) according to its technique parameters as in Figure 1.1, and inspect the screw pitch error with the screw inspector.

Ⅰ. Operation preparation

The preparation of processing trapezoidal thread lead screw is shown in Table 1.2.

Table 1.2　Preparation of turning and inspect lead screw

No.	Name		Preparation
1	Material		Round steel bar ϕ35 mm $\times$ 560 mm
2	Device		CA6140 turning lathe
3	Processing tools	Cutting tools	90° turning cutter, 45° elbow turning cutter, 30° trapezoid thread cutter, A3/8 mm center drill
4		Measuring tools	Vernier caliper 0.02 mm / (0–150 mm, 0–600 mm), micrometer 0.01 mm / (0–25 mm), thread template, lead screw inspector
5		Others	Tool holder, tool setting template, tops(core clamper)

Ⅱ. Operation procedure

The operation steps of trapezoidal thread lead screw are shown in Table 1.3.

Table 1.3　Operation steps of turning and inspect lead screw

No.	Operation procedure	Operation diagram
Step 1	Clamp the workpiece with 20 mm stretching out. 1) Turn the end face; 2) Centering	
Step 2	Clamp the workpiece with 550 mm stretching out and keep the topper still. 1) Use the tool holder to rough and finish turning the outer circle to the size $\phi 28_{-0.25}^{0}$ mm ; 2) Machine undercut to size 10 $\times$ ϕ20 mm at 530 mm; 3) Rough and finish thread to size Tr 28 $\times$ 5	

No.	Operation procedure	Operation diagram
Step 3	Turn round the workpiece and clamp it with the copper gasket	
	1) Ensure that the turning length of end face is 555 mm; 2) Rough and finish turning the outer circle to the size $\phi 18^{-0.018}_{-0.060}$ mm; 3) Chamfer of the workpiece is $C1$ mm.	
Step 4	Test the screw pitch	
	Test the screw pitch error with the screw inspector	

III. Workpiece quality test

The screw thread should be tested according to the technical requirements of the workpiece shown in Figure 1.1.

1. Screw thread and roughness

The large diameter $\phi 28^{0}_{-0.25}$ mm, the middle diameter $\phi 25.5^{-0.052}_{-0.462}$ mm, the small diameter $\phi 22.5^{0}_{-0.565}$ mm should be tested according to the dimensional tolerance, or unqualified if out of tolerance.

The surface roughness of both sides of the thread is $Ra1.6$ μm, or unqualified if degradation.

2. Outer diameter and roughness

The outer diameter dimension $\phi 18^{-0.018}_{-0.060}$ mm should be tested according to the dimensional tolerance, or unqualified if out of tolerance.

The surface roughness of both sides of the thread is $Ra1.6$ μm, or unqualified if degradation.

3. Length

The length dimensions are 530 mm and 555 mm, tested according to the unmarked tolerance GB/T1804—2000.

4. Others

The chamfer Cl mm should be tested according to the unmarked tolerance GB/T1804—2000.

The roughness is $Ra3.2$ μm, or unqualified if degradation.

Ⅳ. Notes

Due to the relatively large diameter of the lead screw, it is of necessity to well cooperate one clamp, one top with the tool holder in processing.

Subtask 2   Turning Long Lead Screw

【Knowledge and Skills Objectives】

(1) Master the method of selecting of cutting parameters for turning long lead screw.

(2) Master the causes and solutions of the deformation of long lead screw.

【Related Knowledge】

Ⅰ. Diagram of long lead screw

The workpiece diagram is shown as in Figure 1.5.

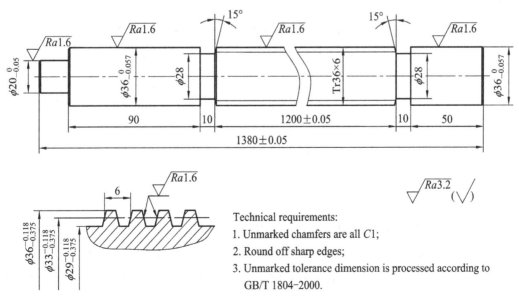

Technical requirements:

1. Unmarked chamfers are all $C1$;

2. Round off sharp edges;

3. Unmarked tolerance dimension is processed according to GB/T 1804—2000.

Figure 1.5 Diagram of long lead screw

The two sides of the long lead screw are smooth shaft heads with strict accuracy requirements, and the middle part of it is the lead screw.

II. Selection of cutting parameters for turning long lead screw

1. Select cutting parameters according to processing methods

In order to remove the excess part of the workpiece as soon as possible in rough turning, a large cutting amount should be selected. In order to ensure the accuracy and surface roughness of the thread during finish turning, a smaller cutting parameter could be selected.

When turning the external thread, the tool rod is short and thick, the rigidity is high and the strength is large, a larger cutting parameters could be selected. When turning the internal thread, the tool rod extends into the hole of the workpiece, and the rigidity and strength are poor, a smaller cutting parameters could be selected.

When turning slender long threads, the rigidity and strength of the workpiece are low, the smaller cutting parameters should be selected. When turning shorter threads, a larger amount of cutting is suitable.

When turning the thread with a small pitch, for every rotation, the relative stroke of the tool on the workpiece is small, and the larger cutting parameters could be selected. When turning the thread with a large pitch, for every rotation, the relative stroke on the workpiece is large, a smaller amount of cutting is necessary.

2. Select cutting parameters according to processing materials

Because brittle materials (such as cast iron) contain more impurities and pores, it is very unfavorable for tool cutting, and an excessively high cutting speed will aggravate tool wear. If the amount of back engagement is too large, it will make the thread tip burst, resulting in waste.

When turning the thread workpiece of plastic material, a large amount of back engagement can be selected accordingly, but the phenomenon of "thrust broadsword" must be prevented.

3. Select cutting parameters according to turning methods

When the direct cutting method is applied, the turning cross section is large, and the stress and heat of the turning tool are serious. It is necessary to select a smaller cutting amount (except for turning threads at high-speed). When the left and right cutting method is adopted, the turning cross section is relatively small, the negative effects of stress and heat from the turning tool are cancelled, and the larger cutting parameters should be selected.

The selection of cutting parameters for the rough turning slender shaft is shown in Table 1.4, and the selection of the finish turning slender shaft is shown in Table 1.5.

Table 1.4   Cutting parameters selection of the rough turning slender shaft (suitable for 75° rough turning cutter)

Diameter of semifinished workpiece D/mm	Cutting parameters		
	Feed amount f/ (mm/r)	Back engagement a_p/mm	Cutting speed v_c/(m/min)
50	0.6	3	60−100
	0.8	4	
	0.6	0.75	
45	0.6	3	60−100
	0.8	4	
	0.6	0.75	
40	0.6	3	60−100
	0.7	3	
	0.6	1.25	
35	0.6	3	60−100
	0.7	3	
	0.6	1.25	
30	0.5	3	40−90
	0.7	2	
	0.5	0.75	
25	0.5	3	40−90
	0.7	2	
	0.5	0.75	
20	0.4	3	40−80
	0.6	2	
	0.4	0.75	

Table 1.5 Cutting parameters selection of the finish turning slender shaft
(suitable for high-speed reverse finish turning steel cutter)

Diameter of semifinished workpiece D/mm	Cutting parameters		
	Feed amount f/ (mm/r)	Back engagement a_p/mm	Cutting speed v_c/(m/min)
50	0.1−0.2	1.5−6	10−15
45	0.1−0.2	1.5−6	10−15
40	0.1−0.2	1.5−6	10−15
35	0.1−0.2	1−4	10−15
30	0.1−0.2	1−4	10−15
25	0.1−0.2	1−4	10−15
20	0.1−0.2	0.7−3	10−15
18	0.1−0.2	0.7−3	10−15
15	0.1−0.2	0.7−3	10−15

Ⅲ. Preventive measures for quality problems when turning long lead screw

The strict motion relationship between the lathe spindle and the cutting tool must be maintained, that is to say, for each rotation of the spindle (i.e., the workpiece rotates one circle), the cutting tool should evenly move the distance of a lead. Their movement relations should be guaranteed as follows: the spindle drives the workpiece to rotate, and the movement of the spindle passes through the hanging wheel to the feed box, the feed box transmits the movement of the spindle to the screw by the variable speed effect (mainly to obtain various pitches); the tool holder driven by the split nuts on the screw and the slide box move linearly, so that the rotation of the workpiece and the movement of the tool are driven by the movement of the spindle, so as to ensure the strict motion relationship between the workpiece and the tool. In the actual turning, due to various reasons, if the movement between the spindle and the tool has problems in a certain link, it would cause the failure of the turning thread, especially the failure in turning the long lead screw, which should be solved in time. Common quality problems and solutions in turning threads are shown in Table 1.6.

Table 1.6   Common quality problems and solutions in turning threads

No.	Quality problems	Reasons	Analysis	Solutions
1	Blade jamming	Tools are installed too high or too low, so workpiece is not clamped properly or tool wear is too severe	When the lathe tool is installed too high, if the back engagement reaches a certain value, the rear of the lathe tool holds the workpiece, which will increase the friction force, and even bend the workpiece to the top, resulting in the phenomenon of blade jamming. When the tool is installed too low, the chip is not easy to discharge, the direction of the radial force of the tool is in the center of the workpiece, and the gap between the horizontal screw and the nut is too large, which automatically deepens the back engagement, so that the workpiece is elevated and the phenomenon of blade jamming occurs	The tool height should be adjusted in time so that the tool tip is equal to the axis of the workpiece (the tool can be countered at the top of the tailstock)) and adjust the clearance of the lateral feeding lead screw nut
		The workpiece is not clamped properly	The rigidity of the workpiece itself cannot bear the cutting force when cutting, resulting in excessive deflection, which changes the center height of the lathe tool and the workpiece (that is to say, the workpiece is elevated), resulting in a sharp increase in the cutting depth and blade jamming phenomenon	The workpiece should be fixed firmly with the top of the tailstock so as to increase the rigidity of the workpiece
		The tool wear is too severe	It makes the cutting force increased, which bends the workpiece, resulting in blade jamming phenomenon	The tool should be repaired and grinded

No.	Quality problems	Reasons	Analysis	Solutions
2	Disorderly buckles	When the lathe screw rotates one round, the workpiece does not correspondingly rotate a full round	When the ratio of the screw pitch of the lathe to the screw pitch of the workpiece is not an integral multiple, if the slit nuts are open and the saddle is shaken to the starting position when the tool is retracted, then when the slit nuts are closed again, the tip of the lathe will not be in the spiral groove pulled out by the previous cutting, resulting in disorderly buckles	When turning the long screw, the forward and backward turning method is used to retract the tool. As the transmission between the spindle, screw, and tool holder has not been separated, and the turning tool is always in the original spiral groove, then there are no disorderly buckles. When rough-turning the long screw, if the machine tool lead can be an integral multiple of the part lead, lift the opening and closing nut, and perform rough turning, open the slit nuts, and manually process tool retracting, the tool retracting speed should be quick, which is beneficial for improving productivity and maintaining lead screw accuracy
3	Incorrect screw pitches	The pitch is incorrect on the full length of the thread	Due to the axis endplay of the lathe screw	The round nut at the joint of the lathe screw and feed box should be adjusted to eliminate axial clearance of the thrust ball bearing at the joint.
			Due to the axis endplay of the lathe spindle	Adjusting nut should be adjusted to eliminate axial clearance of back thrust ball bearing.
			The slit nuts of the slide box and the screw are not on the same axis, resulting in poor meshing	The slit nuts and their clearance should be repaired and adjusted
			The wear of the slit nuts in the slide box causes the instability of the slit nuts	Swallow-tail guideway and insert strip should be prepared to meet the matching requirements
			The hanging wheel gap is too large	The hanging wheel clearance should be readjusted

Continued Table 2

No.	Quality problems	Reasons	Analysis	Solutions
4	The appearance of bamboo lines	There is a periodic error in gear transmission from spindle to screw	The low rotation center of the gear in the hanging wheel box and the gear in the feed box is caused by its own manufacturing error, some localized wear or eccentric installation of the gear on the shaft. The low rotation causes the periodic uneven rotation of the screw and the periodic uneven movement of the tool, resulting in the occurrence of bamboo lines	Gears with errors or wear should be repaired
		There is an error in the following rest	The following rest is not stable, loose or tight	The support clearance of the following rest should be adjusted
5	Incorrect medium diameter	It may be caused by the deep penetration of the cutting tool, inaccurate dial plate and not timely measurement	The dial plate is loose and the allowance is not appropriate in fine turning	Blade edge of the cutting tool should keep sharp, and the reading should be timely measured
6	Rough thread surface	It is caused by unsmooth blade edge of the cutting tool, inappropriate cutting fluid used, inappropriate cutting speed and workpiece material, and vibration during cutting process	The surface roughness increases when turning thread, not only for the reason of equipment, but also of tools, operators and so on. When excluding quality problems, we should make specific analysis, find out specific factors through various detection and diagnosis methods, and take effective solutions	Correct adjust grinding wheel or cutting tool grinded with oilstone is of necessity; appropriate cutting speed and cutting fluid should be selected; the insert strip of saddle platen, and middle and small slide tail guide rail of lathe bed should be adjusted to ensure the accuracy of lead rail clearance and prevent vibration during cutting

【Skills Training】

Turning Long Lead Screw

Processing long lead screw according to its technique parameters in Figure 1.5.

Ⅰ. Operation preparation

The preparation of processing long lead screw is shown in Table 1.7.

Table 1.7   The preparation of processing long lead screw

No.	Name		Preparation
1	Material		45 steel ϕ40mm×1450mm
2	Device		CA6140 turning lathe
3	Processing tools	Cutting tools	90° turning cutter, 45° elbow turning cutter, 30° trapezoid thread cutter, grooving cutter, A3.15/8 mm center drill
4		Measuring tools	Vernier caliper 0.02 mm / (0–150 mm, 0–600 mm), micrometer 0.01 mm / (0–25 mm), thread template, a lead screw inspector
5		Others	Follow rest, center frame, drill clip, movable wrench, fixed top (core clamper), movable top

Ⅱ. Operation procedure

The operation steps of processing long lead screw is shown in Table 1.8.

Table 1.8   Operation steps of processing long lead screw

No.	Operation procedure	Operation diagram
Step 1	Clamp the workpiece tightly with 1430 mm stretch out; align and support the other end with a center frame	
	Turning the end face and drill the center hole	
	1) Remove the center frame and install the tool holder; 2) Rough-turning the outer circle ϕ38 mm with the length of 60 mm; 3) The chamfer is C1 mm.	

Continued Table

No.	Operation procedure	Operation diagram
Step 2	U-turning the clamper, align and erect the center frame 1) Turning the total length of 1380 ± 0.5 mm; 2) Drill the center hole and clamp the workpiece by the core clamper; 3) Adjust the center frame at a distance of 70 mm away from the end face, turning the escrape, for the size $\phi 10$ mm $\times \phi 28$ mm, at a distance of 500 mm away from the end face, the chamfer is 15°. 4) Turning the escrape on the other side for the size $\phi 10$ mm $\times \phi 28$ mm, with guaranteed length (1200 ± 0.5) mm; 5) The chamfer is 15°	
Step 3	Set up the follow-rest 1) Rough turning the thread at Tr36 × 6; 2) Finish turning the thread at Tr36 × 6; 3) After turning the thread, blunt the edges of the tooth top	
Step 4	Adjust the center frame at a distance of 1000 mm away from the end face 1) Finish turning the outer circles at the two sides for $\phi 36_{-0.057}^{0}$ mm, $Ra \leqslant 1.6$ μm; 2) Cut the workpiece and ensure its length of 1381 mm; 3) Turning end face and ensure its length of 1380 mm; 4) The chamfer is $C1$ mm	
Step 5	Turning the workpiece around and clamp it with a copper pad 1) Drill the center hole and support the top; 2) Remove the center frame, put the follow-rest and turning the outer circle $\phi 20_{-0.05}^{0}$ mm, $Ra \leqslant 1.6$ μm with the length of 20 mm; 3) The chamfer is $C1$ mm	

III. Workpiece quality test

The workpiece should be tested according to the technical requirements of the workpiece shown in Figure 1.5.

1. Screw thread

The large diameter $\phi 36_{-0.375}^{-0.118}$ mm, the medium diameter $\phi 33_{-0.375}^{-0.118}$ mm and the small diameter $\phi 29_{-0.475}^{-0.118}$ mm should be tested accurately according to the dimensional accuracy, or unqualified if out of tolerance.

Surface roughness is $Ra1.6$ μm, or unqualified if there is downgrading.

2. Outer diameter and roughness

The external diameter $\phi 36_{-0.057}^{0}$ mm and $\phi 20_{-0.05}^{0}$ mm are all tested according to dimensional tolerance, or unqualified if out of tolerance. Roughness is $Ra1.6$ μm, or unqualified if there is downgrading.

3. Length dimensions

The length dimensions of 1200 ± 0.05 mm and 1380 ± 0.5 mm are tested according to the dimensional tolerance, or unqualified if out of tolerance.

The length dimensions of 90 mm, 10 mm, 10 mm, and 50 mm with unspecified tolerance are all tested according to dimensional tolerance of GB/T1804—2000.

4. Other parts

The chamfer $C1$ mm is processed according to unspecified dimensional tolerance of GB/T 1804—2000. Other surface roughness is $Ra3.2$ μm, or unqualified if there is downgrading.

IV. Notes

(1) When turning the outer circle, properly adjust the contact pressure between the tool holder and the workpiece at any time.

(2) Detect whether the axis of the tailstock sleeve is coaxial with the axis of the spindle.

(3) After the top support the workpiece, the tension must be appropriate.

(4) If the size of the threaded part is the main part of the workpiece, and the size of the shaft at both ends is relatively high, first semi-finish turning the shaft, then turning the thread, and finally finish turning the thread.

(5) When the workpiece is bent, it must be corrected before processing.

Task 2   Machining Multi-start Thread and Multi-line Worms

Subtask 1   Machining Multi-start Thread

【Knowledge and Skills Objectives】

(1) Master the dividing method of multi-start thread.

(2) Master the solutions to common problems when turning multi-start thread.

【Related Knowledge】

Ⅰ. Workpiece diagram recognition of multi-thread trapezoidal thread

The diagram of three-start conical trapezoidal thread (one type of multi-start trapezoidal thread) is displayed in Figure 1.6.

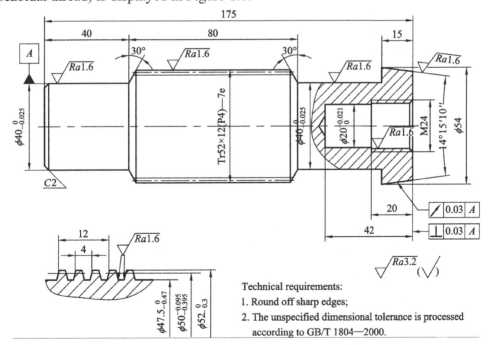

Technical requirements:
1. Round off sharp edges;
2. The unspecified dimensional tolerance is processed according to GB/T 1804—2000.

Figure 1.6   The diagram of three-start conical trapezoidal thread

The right end of the part is a cone head sample, and the middle is a three-start trapezoidal thread with a pitch of 4 mm and a lead of 12 mm. The thread tolerance accuracy is 7e, the thread has tolerance requirements for large, medium and small diameters, and the surface roughness on both sides of the thread is $Ra1.6\,\mu m$. In the outer circle, there are sizes in 2 places as $\phi40^{\ 0}_{-0.025}\,mm$, and the left end $\phi40^{\ 0}_{-0.025}\,mm$ is the reference axis. The surface roughness is $Ra1.6\,\mu m$. The conical shaft diameter of the big end in the right end is $\phi54\,mm$, the angle of cone head is required as in Figure 1.7. The inner hole diameter is $\phi20^{+0.021}_{\ 0}\,mm$.

II. The method of dividing lines

The spirals of the multi-start thread (or worm) are equally spaced along the axial direction. From the distribution of spirals end faces (see Figure 1.7), it can be seen that the starting points of the spirals are at the same angle on the circumference. In the turning process, solving the problem of equidistant distribution of spirals is called dividing or splitting. If the isometric error is too large, it will affect the matching accuracy of the internal and external threads and the meshing accuracy of the worm and the worm wheel, and reduces the workpiece's service life.

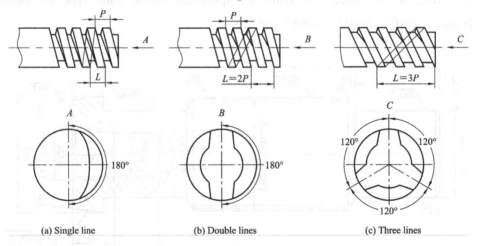

(a) Single line	(b) Double lines	(c) Three lines

Figure 1.7   The distribution of spirals on end face

According to the characteristics of the equidistant distribution of the helical lines of the multi-start thread (or worm) in the axial and circumferential directions, there are two types of splitting methods: axial dividing method and circumferential dividing method.

1. Axial dividing method

Axial dividing method is to move the turning tool along the thread (or worm) axis by a pitch (circumferential section) after turning the first spiral groove, then turning the second spiral groove. This method only requires that the movement of the turning tool along the axial is accurately controlled. The specific applications of this method are as follows.

1) Small skateboard scale dividing method

First, the guide rail of the small skateboard should be aligned to be parallel to the axis of the spindle. After turning the first spiral groove, the small skateboard should be moved forward or backward a pitch (or circumferential section), then turning the second spiral groove. The moving distance of the small skateboard can be controlled by the scale of the small skateboard, and the number of grids, K, can be calculated by the following formula:

$$K = \frac{P}{a} \tag{1.3}$$

here, P: workpiece pitch or worm circumference, unit mm; a: the moving distance of each grid of the small skateboard, unit mm.

Example 1.1   When turning the Tr40 × 14 (P7) thread, the scale of the small skateboard of the lathe is 0.05 mm. Please find the number of grids that the small skateboard should turn when dividing the line.

Solution: if $P = 7$ mm, $a = 0.05$ mm,

$$K = \frac{P}{a} = \frac{7}{0.05} = 140 \text{ grids}$$

It is easier to use a small skateboard to divide the scale without other auxiliary tools, but the isometric accuracy is not high.

2) Dial indicator and gauge block dividing method

When dividing parts with high requirements for isometric accuracy, dial indicator and gauge block dividing method can be used to control the moving distance of the small skateboard, as shown in Figure 1.8. The dial indicator is clamped on the tool holder, and a stop on the saddle (slider) is fastened. Before the first spiral groove, adjust the small skateboard to make the contact of the dial indicator contact with the stopper, set the dial indicator to be zero. When the first spiral groove is completed, move the small skateboard, and the reading indicated by the dial indicator is the distance that the small skateboard moves. When dividing a multi-start thread (or worm) with a larger pitch (or circumferential pitch), due to the limit of the dial indicator, a gauge block (or a group) can be inserted between the dial indicator and the stopper. The thickness of the gauge block is preferably equal to the pitch (or circumferential pitch) of the workpiece.

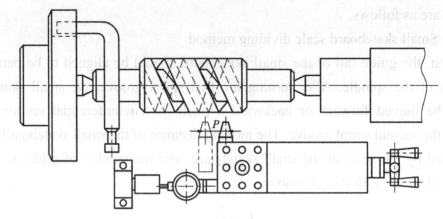

Figure 1.8　Dial indicator and gauge block dividing method

The accuracy of this dividing method is high, but because of the vibration during turning, it is easy to move the dial indicator. It must be guaranteed that the indicator is aligned to the zero position.

2. Circumferential dividing method

The circumferential dividing method is based on the characteristics of the isometric distribution of the helix on the circumference, when the first spiral groove is turned, the transmission chain between the workpiece and the screw is removed, the workpiece is turned over an angle $\theta \left(\theta = \dfrac{360°}{n} \right)$, then the transmission chain between the workpiece and the screw is connected, and the other spiral groove is turned, so that the division is completed in turn.

The specific applications of circumferential dividing method are as follows.

1) Exchange gears dividing method

In general, the rotation speed of the exchange gear Z_1 is equal to the rotation speed of the spindle, and the rotation angle of Z_1 is equal to the rotation angle of the workpiece. Therefore, when the number of teeth of Z_1 is an integral multiple of the number of threads or worms, this method can be applied.

The schematic diagram of the exchange gear dividing method is shown in Fig. 1.9. When a spiral groove is ready, stop and cut off the power supply, equalize the spiral groove on Z_1 according to the number of lines, mark a and d with chalk at the meshing point with Z_1, for example, the tooth number of Z_1 is 60, when turning the three-line thread, mark b and c at the tooth number of 20 from the mark a, and then loosen the exchange gear rack, so that Z_1 and Z_2 are divided. Rotate the spindle by hand, so that the mark b or c is aligned with the mark d, and then mesh Z_2 and Z_1, so the second spiral groove can be turned. The same method is used before turning the third spiral.

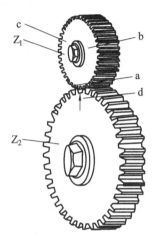

Figure 1.9　Schematic diagram of exchange gear dividing method

The advantage of this method is the high accuracy of thread dividing, but the number of threads or worms to be turned is limited by the number of teeth Z_1, and the operation is more troublesome, so it is rarely used in mass production.

2) Chuck claw dividing method

When the workpiece is clamped between two centers, for example, when using a chuck instead of a dial, the jaws can be used to divide 2, 3, and 4 threads or worms. When dividing, it only needs to loosen the back core clamper, rotate the workpiece together with the heart carrier to an angle by another jaw on the chuck, and then press the back centre to turn another spiral groove.

This method is relatively simple, but the accuracy is not high.

3) Index plate dividing method

As shown in Figure 1.10, it is the schematic diagram of index plate dividing method used for turning multi-start threads (or worms). Installed on the main shaft of the lathe, the turntable 4 has high accuracy positioning holes 2 (usually designed as 12 holes or 24 holes), which can be used for dividing 2, 3, 4, 6, 8 and 12 start threads or worms.

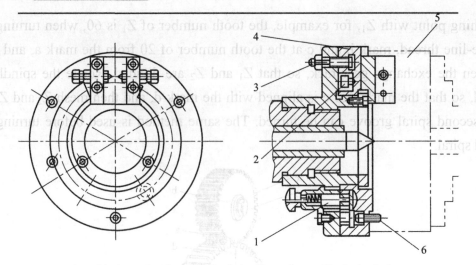

1-positioning socket; 2-positioning holes; 3-nut; 4-turntable; 5-chuck; 6-screws.

Figure 1.10   Schematic diagram of the index plate dividing method

When dividing, first stop turning, loosen the nut 3, pull out the positioning socket 1, rotate the turntable 4 to an angle of θ, and then insert the positioning socket into another positioning holes, tighten the nut, and the dividing is completed.

This dividing method has the advantages of high precision and easy operation, which is an ideal dividing method.

Ⅲ. Common problems and solutions when turning internal threads

When turning the internal threads, sometimes although the thread depth has been turned to the required size, it still cannot be screwed in when using a thread plug gauge or checking the matching external thread; sometimes, although it can be screwed in, the fit is too loose if it is screwed several times, or just one end can be screwed in and the other end cannot; sometimes, only a few teeth are screwed in at the thread entrance.

1. Problems and reasons

(1) The two sides of the turning tool are not straight, so that the two sides of the thread profile by turning are correspondingly not straight, which reduces the thread accuracy.

(2) The top width of the turning tool is too narrow. Although the thread has been turned to the specified depth, the space width at the pitch diameter of the internal thread does not reach the required size(it is smaller than the space width at the pitch diameter of the external thread).

(3) Due to the turning tool for having inaccurate sharpening blade, or

excessively large front angle, or too high or too low, a large thread angle error occurs which would reduce the thread accuracy. Or because the turning tool is not installed straightly, a large thread half-angle error is produced, and the thread angle is correspondingly skewed. During inspection, one end would be screwed in, the other end would not be screwed in, or the fit would be too loose.

(4) The bottom diameter of the internal thread is too small, and the thread cannot be screwed in during inspection.

(5) The shank of the turning tool cannot be made too thick or too short due to the limitation of the aperture size and length, so the rigidity is poor, and it will produce a small amount of bending deformation due to the influence of the cutting force during turning. The commonly known "cutter back-off" phenomenon causes tooth profile errors in the internal thread, so only a few teeth can be screwed in at the entrance during inspection.

2. Solutions

Different methods are adopted to solve different problems when turning internal threads. At the same time, "prevention first" is advocated.

(1) The turning tool should be sharpened and installed correctly. When using the straight cutting method, it must be noted that the width of the tip of the thread turning tool should meet the requirements, so that after the internal thread is turned to the specified depth, the width of the tooth profile at the middle diameter basically reaches the specified size. The top width of the trapezoidal internal thread turning tool needs to be reduced by 0.05−0.1mm according to the calculated top width.

(2) When using the straight cutting method, if the internal thread has been turned to the specified depth, but the space width has not yet reached the required size, move the small skateboard to perform single-sided turning at the original back-engagement position (usually by making the small slide move toward the end of the bed) until the external thread is screwed in.

(3) When the machining volume is large, the turning tool must be sharpened again to make the top width of the turning tool meet the requirements before turning.

(4) When turning the bottom diameter hole of the internal thread, the required size must be ensured.

(5) For the thread taper error caused by the phenomenon of "cutter back-off", don't blindly increase the amount of back-engagement, otherwise, not only the (cone)

error could not be reduced, but also the fitting accuracy of the thread would be affected. At this time, the method of "sliding the knife" is to make the turning tool work repeatedly at the original position of the back of the tool, in order to gradually eliminate the tapering error, and then use the external thread to test until it is fully screwed in.

IV. Machining process and calculation

When machining the three-start conical trapezoidal thread shown in Figure 1.6, firstly divide the line on the surface of the outer circle with the tool nose, check the thread lead, and then perform turning. When turning a cone, the technical requirements are a circular runout of 0.03 mm and a perpendicularity of 0.03mm relative to the reference A. When machining a cone and the outer circle $\phi 40_{-0.025}^{0}$ mm, use two sharp points to support the two end faces, ensure that the technical requirements of the cone are met.

The formula for measuring the three-start conical trapezoidal thread by using the three needles method is as follows:

$$M = d_2 + d_D \left(1 + \frac{1}{\sin \frac{\alpha}{2}} \right) - \frac{P}{2} \cot \frac{\alpha}{2}$$

$$= d_2 + 4.864 d_D - 1.866 P \tag{1.4}$$

Measuring needle diameter $d_D = 2.072$ mm, then

$$M = 50 + 4.864 \times 2.072 - 1.8664 \times 4$$
$$= 52.61 \text{ mm}$$

【Skills Training】

Turning Multi-start Trapezoidal Thread

I. Operation preparation

The preparation for turning one type of the multi-start trapezoidal thread (its diagram is shown in Figure 1.6) is shown in Table 1.9.

Table 1.9   The preparation for turning one type of the multi-start trapezoidal thread

No	Name		Preparation
1	Material		45 steel, ϕ40 mm × 1450 mm
2	Device		CA6140 turning lathe
3	Processing tools	Cutting tools	90° turning cutter, 90° anti-deviation turning cutter, 45° turning cutter, 30° trapezoid thread cutter, grooving cutter, external fine turning tool, ϕ18 mm drill, inner hole cutter, M24top, ϕ20H7reamer, A2/5 mm center drill
4		Measuring tools	Universal protractor 2′(0° −320°), vernier caliper 0.02 mm / (0−200 mm), micrometer 0.01 mm / (0−25 mm, 25−50 mm, 50−75 mm), 30° angle template, vernier depth gauge 0.02 mm / (0−200 mm) and ϕ2.072 mm gauge pin
5		Others	Screwdriver for slotted head screws, adjustable wrench, core clamper, drill fixture, other common tools

II . Operation procedure

The operation steps of turning multi-start trapezoidal thread is shown in Table 1.10.

Table 1.10   Operation steps of turning multi-start trapezoidal thread

No.	Operation procedure	Operation diagram
Step 1	The outer circle is clamped with 145mm stretch out, turning the end face 1) Drill the center hole and hold the workpiece tightly with a core clamper; 2) Rough and fine turning the outer circle $\phi40_{-0.025}^{0}$ mm and with a length of 40 mm; 3) Rough and fine turning the outer circle $\phi52_{-0.30}^{0}$ mm and with a length of 85.5 mm; 4) Rough and fine turning the undercut groove $\phi40_{-0.025}^{0}$ mm and with a length of 80 mm; 5) The chamfer at each of two ends is 30°	

No.	Operation procedure	Operation diagram
Step 2	Turn the workpiece around and clamp it tightly with the copper gasket 1) Turning the end face to ensure the total length of 175 mm; 2) Drill the center hole and clamp it tightly with the two tops; 3) Dividing lines on the outer circle surface with turning cutter to separate out three spiral lines, rough and fine turning the trapezoidal thread to the size $\phi 52_{-0.30}^{0}$ mm, and make the sharp corner of the tooth tip blunt; 4) Finish turning both ends to size $\phi 40_{-0.025}^{0}$ mm ; 5) The chamfer is $C2$ mm; 6) Fine-turning the face of the outer cone $14°15'$ and the diameter of the outer cone to size $\phi 54$ mm	
Step 3	Remove the workpiece and clamp it with copper gasket of the thread 1) Drill the hole of $\phi 18$ mm, length 44 mm; 2) Ream the hole to $\phi 19.7$ mm ; 3) Ream inner hole to size $\phi 20_{0}^{+0.021}$ mm ; 4) Turn the inner hole to $\phi 21$ mm and with a length of 20 mm. 5) Tap thread M24	

III. Workpiece quality test

The workpiece quality test shall be carried out according to the technical requirements of the multi-start trapezoidal thread as shown in Figure 1.6.

1. The three-line(three-start) trapezoidal thread and roughness

The major diameter $\phi 52_{-0.30}^{0}$ mm should be tested according to the dimensional tolerance, unqualified if out of tolerance.

The middle diameter $50_{-0.395}^{-0.095}$ mm (3 places) should be tested according to the dimensional tolerance, unqualified if out of tolerance.

The minor diameter $\phi 47.5^{0}_{-0.47}$ mm (3 places) should be tested according to the dimensional tolerance, unqualified if out of tolerance.

The roughness of both sides and top surface of the thread is Ra 1.6 μm, unqualified if downgrading.

2. Outer diameter, inner diameter and roughness

The outer diameter $\phi 40^{0}_{-0.025}$ mm (2 places) should be tested according to the dimensional tolerance, unqualified if out of tolerance.

The inner hole $\phi 20^{+0.021}_{0}$ mm should be tested according to the dimensional tolerance, unqualified if out of tolerance.

The roughness is Ra 1.6 μm, unqualified if downgrading.

3. Geometric tolerances and cone angle

The circular runout of 0.03 mm and the verticality of 0.03 mm are tested according to the shape tolerance and position tolerance, unqualified if out of tolerance.

The unspecified tolerance of cone angle of 14° 15′10″ shall be tested according to GB/T1804—2000.

4. Length dimension

The unspecified tolerance of lengths of 40 mm, 80 mm, 175 mm, 15 mm, 42 mm, 20 mm shall be tested according to the standard of GB/T 1804—2000.

5. Thread

The M24 thread is unqualified if the thread form is incomplete.

6. Others

Other roughness is Ra 3.2 μm, unqualified if downgrading.

Subtask 2   Machining Multi-line Worms

【Knowledge and Skills Objectives】

(1) Master the basic knowledge of multi-line worms.

(2) Master the calculation method for the dimensions of thread form of worms.

(3) Master the turning method of multi-line worms.

(4) Master the method of grinding the cutting tool of worms.

【Related Knowledge 】

Ⅰ. Diagram of multi-line worms and shaft

The diagram of single crank multi-line worms and shaft (one type of multi-line worms and shaft)is shown as in Figure 1.11.

(1) There is a step shaft of $\phi 35_{-0.025}^{0}$ mm and 20 mm in length on the left end of the workpiece.

(2) The diameter of the eccentric shaft is $\phi 25_{-0.021}^{0}$ mm and the length of it is $\phi 20_{0}^{+0.05}$ mm .

(3) The parameters of right-end worms are modulus 3, line number 4, tooth thickness $4.55_{-0.078}^{-0.025}$ mm.

(4) The right end shaft diameter is $\phi 35_{-0.025}^{0}$ mm, the right end is with the length of 20mm.

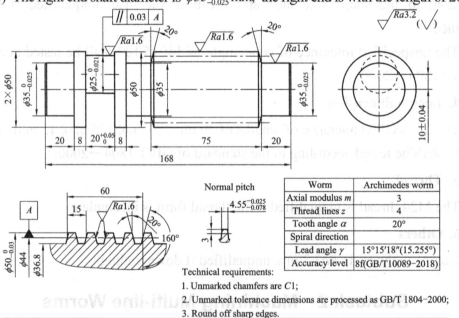

Worm	Archimedes worm
Axial modulus m	3
Thread lines z	4
Tooth angle α	20°
Spiral direction	
Lead angle γ	15°15′18″(15.255°)
Accuracy level	8f(GB/T10089-2018)

Technical requirements:

1. Unmarked chamfers are C1;

2. Unmarked tolerance dimensions are processed as GB/T 1804-2000;

3. Round off sharp edges.

Figure 1.11　Diagram of single crank multi-line worms and shaft

Ⅱ. Basic knowledge of multi-line worms

Thread and worms have single or multiple line(s), as shown in Figure 1.12. The thread formed along one helix is called single thread, and the thread formed along two or more helixes with axial equidistant distribution is called multi-line thread (multi-line worms).

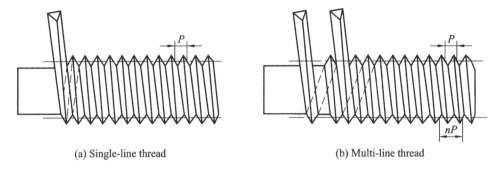

(a) Single-line thread   (b) Multi-line thread

Figure 1.12   Threads

The pitch (expressed by P) of a single-line or multi-line thread refers to the axial distance between two adjacent teeth on the same helix. The lead of multi-line thread(thread lead)refers to the distance between the corresponding two points on the adjoining pitch lines. The lead of multi-line thread(expressed by L) is equal to the number of lines multiplied by the pitch, that is, $L = nP$.

The circular pitch of multi-line worms refers to the arc distance between two adjacent teeth on the same helix. It is marked by P_x, $P_x = \pi m_x$.

The axial pitch of multi-line worms refers to axial distance between the corresponding two points on the diameter line of the indexing circle, noted by P_z, it is equal to the number of knots multiplied by the number of lines, i.e. $P_z = \pi m_x z_1$, z_1 is the lines of the worms.

The lead of the multi-line thread (worms) is greater than the pitch (circumferential pitch) of it. When calculating the thread lift angle and the lead angle of the worms, the lead must be calculated, namely

$$\tan\varphi = \frac{nP}{(\pi d_2)} \tag{1.5}$$

$$\tan\gamma = \frac{P_z}{(\pi d_1)} \tag{1.6}$$

In the above formulas, φ, thread lift angle; nP, thread lead, n is the number of lines; d_2, thread pitch diameter; γ, lead angle; d_1, the diameter of the pitch circle.

The calculation method of the size of each part of the multi-line thread(or multi-line worms) is the same as that of the single-line thread.

III. Calculation of dimensions for various parts of multi–line worms

Worms in metric and imperial units are available. The multi-line worms is the commonly-used device in the market. The dimensions of each part of the multi-line worms are calculated in Table 1.11.

Table 1.11   Calculation of different parts of the multi-line worms

Name	Code	Formula
Axial modulus	m_x	Basic parameter
Tooth angle (Pressure angle)	α	$\alpha = 20°$
Circular pitch	P_x	$P_x = \pi m_x$
Axial pitch	P_z	$P_z = z_1 P_x = z_1 \pi m_x$
Full tooth height	h	$h = 2.2 m_x$
Tooth top	h_{a1}	$h_{a1} = m_x$
Tooth root height	h_{f1}	$h_{f1} = 1.2 m_x$
Pitch circle diameter	d_1	$d_1 = d_{a1} - 2 m_x$
Tooth tip diameter	d_{a1}	$d_{a1} = d_1 + 2 m_x$
Tooth root diameter	d_{f1}	$d_{f1} = d_1 - 2.4 m_x$ or $d_{f1} = d_{a1} - 4.4 m_x$
Tooth top width	s_a	$s_a = 0.843 m_x$
Axial tooth root groove width	W_f	$W_f = 0.697 m_x$
Tooth root groove width in normal direction	W_{fn}	$W_{fn} = 0.697 m_x \cos\gamma$
Lead angle	γ	$\tan\gamma = \dfrac{P_z}{\pi d_1} = \dfrac{m_x z_1}{d_1}$
Axial tooth thickness	s_x	$s_x = \dfrac{P_x}{2}$
Normal tooth thickness	s_n	$s_n = \dfrac{P_x}{2}\cos\gamma$

IV. Turning methods and steps for multi-line worms

1. Reasonable choice of turning methods

Due to the large lead of the multi-line worms, low-speed cutting is generally used. The turning should be divided into two stages: rough turning and finish turning.

When rough turning the tooth profile, the appropriate cutting method and the feed method should be selected.

In order to prevent "breaking edges" caused by processing three edges cutting at the same time during rough turning, the left and right cutting method can generally be used. When the multi-line worms with the modulus $m_x > 3$ mm is used, the turning tool with a width smaller than the root groove width of the worms should be used first, then turning the worms to the size of the root circle diameter; when rough turning the multi-line worms with the modulus $m_x > 5$ mm, the layered cutting method can be used to reduce the cutting area of the turning tool and make the cutting go smoothly.

When finish turning the tooth profile, a fine turning tool with a chip groove should be used to turn the tooth flank by the left and right cutting method.

2. Turning steps of multi-line worms

When turning a multi-line worms, it is not possible to turn one spiral groove completely before turning another spiral groove. In order to ensure the quality in machining worms, the following steps can be used during turning, the diagram of turning sequence for three-line worm by axial dividing method is shown as in Figure 1.13.

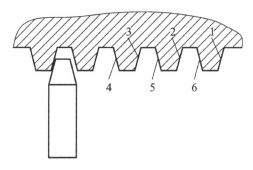

Figure 1.13   Diagram of turning sequence for three-line worms by axial dividing method

(1) After rough turning the first spiral groove, the scale of the middle and small sliding plates should be noted and kept in mind.

(2) According to the accuracy of the tooth profile, dividing the line and rough turning the second and third spiral grooves. If the circumferential dividing method is used, the scale of the middle and small sliding plates should be the same as the scale of the first spiral groove; if the axial dividing method is used, the scale of the middle sliding plate should be the same as the scale of the first spiral groove, and the small sliding plate moves exactly one pitch.

(3) Finish turning each spiral groove according to the above method until it meets the requirements. In addition, when using the axial dividing method together with the left and right cutting method for finish turning, in order to ensure the accuracy of the axial tooth pitch, it is required that the left and right movement of the turning tool (that is, the amount of borrowing) when turning the spiral groove should be equal, and the turning sequence of each tooth flank of multi-line worms should also be noted, that is to say, first turning each tooth flank in the same direction one by one, then turning each tooth flank in the other direction one by one. Otherwise, the accuracy of the axial tooth pitch will be affected due to the error caused by the clearance of the screw and the nut of the small sliding plate.

3. Modification of the unevenness of the split ends of the multi-line worms

As shown in Figure 1.14, according to the split ends of the double-line worms, the axial pitch of the worms is $P_z = a + b + a' + b'$ and the circular pitch of the worms is $P_x = \dfrac{P_z}{2} = a + b = a' + b'$. There are the following situations when measuring with a single needle or with a tooth thickness caliper.

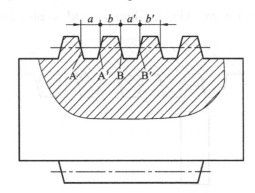

Figure 1.14　Modification of the unevenness of the split ends of the double-line worms

(1) $a = a'$, $b > b'$. It is necessary to evenly modify the sides A' and B of the tooth flake to make $a = a'$, $b = b'$.

(2) $b = b'$, $a > a'$. It is necessary to evenly modify sides B and B' of the tooth flank to make $a = a'$, $b = b'$.

(3) $a > a'$, $b < b'$. It is necessary to modify the side B' of the tooth flank to make $a = a'$, $b = b'$.

(4) $a = a'$ (or $b = b'$). When the size is at the lower deviation limit, since the four tooth flank sides of A, A', B and B' have no modification margins, any

modification of the unevenness of the dividing line will produce over tolerance of b or b'. Therefore, the unevenness of the dividing line could only be modified when the tooth thickness b or b' of the tooth flank has a margin.

V. Turning tools for grinding worms

The cutter head of the worms can be divided into rough turning head and finish turning head, which are grinded by ordinary high-speed steel. The blade angle of the right-handed rough turning tool is shown in Figure 1.15. The basic size of the tooth profile of the right-handed rough turning tool is shown in Figure 1.16.

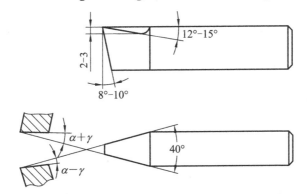

Figure 1.15   The blade angle of the right-handed rough turning tool

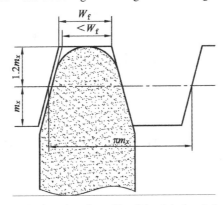

Figure 1.16   The basic size of tooth profile of the right-handed rough turning tool

During rough turning, the width of the tip of the worms in Fig 1.16 should be smaller over 1 mm than the width of the groove bottom of the worms. When grinding the worms, the fine turning allowance and the left and right turning allowance should be taken into account. During semi-finishing turning, the tip of the worms should be smaller about 0.5 mm than W_f. The process of manufacturing the tip arc is as follows: first grind the width of the tip to a certain size, then use a grinding wheel to grind the entire tip into a circular curved blade, which can increase the life of the rough turning tool.

The basic size of the tooth profile of the right-handed fine turning tool is shown in Figure 1.17.

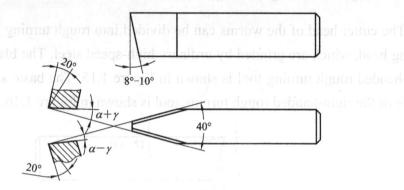

Figure 1.17　The size of the tooth profile of the right-handed fine turning tool

The Archimedes worms is processed by form turning, and the shape of the cutting edge of the fine turning tool is exactly the same as the axial tooth profile of the worms, so the size and the tooth profile angle of the fine turning tool are grinded according to the axial tooth profile of the worms. When installing, it should be guaranteed that the cutting edges on both sides are on the rotation center of the workpiece, which aims to ensure the accuracy of the tooth profile of worms. When grinding, the two edges must be straight and sharp. The surface roughness of the front and the back of the edges of the turning tool should be Ra 0.4 μm or less, so it must be carefully grinded.

VI. Processing analysis and size calculation of multi-line worms

1. Turning process

(1) The workpiece adopts the one clamping and one top method. First, turning the middle of worms to size $\phi 35$ mm, turning the right end face to size $\phi 35_{-0.025}^{0}$ mm. Next, rough and fine turning worms according to dividing line, turn around and clamp the workpiece, turning the outer circle to sizes of $\phi 50$ mm, $\phi 35_{-0.025}^{0}$ mm and the crank of $\phi 25_{-0.021}^{0}$ mm.

(2) Before turning the worms, it needs to draw a spiral line on the outer surface and measure the tooth pitch with a caliper.

2. Calculation of the size of each part of the multi-line worms

The calculation of each part of the multi-line worms is shown in Table 1.12.

Table 1.12 The calculation of the left-hand multi-line worms

No.	Name	Formula
1	Circular pitch	$P_x = \pi m_x = 3.14 \times 3 = 9.42$ mm
2	Axial pitch	$P_z = z_1 P_x = 4 \times 9.42 = 37.68$ mm
3	Full tooth height	$h = 2.2 m_x = 2.2 \times 3 = 6.6$ mm
4	Tooth top	$h_{a1} = m_x = 3$ mm
5	Tooth root height	$h_{f1} = 1.2 m_x = 1.2 \times 3 = 3.6$ mm
6	Pitch circle diameter	$d_1 = d_{a1} - 2m_x = 50 - 2 \times 3 = 44$ mm
7	Tooth tip diameter	$d_{a1} = d_1 + 2m_x = 44 + 2 \times 3 = 50$ mm
8	Tooth root diameter	$d_{f1} = d_1 - 2.4 m_x = 44 - 2.4 \times 3 = 36.8$ mm
9	Tooth top width	$s_a = 0.843 m_x = 0.843 \times 3 = 2.529$ mm
10	Axial tooth root groove width	$W_f = 0.697 m_x = 0.697 \times 3 = 2.091$ mm
11	Lead angle	$\tan\gamma = \dfrac{P_z}{\pi d_1} = \dfrac{37.68}{3.14 \times 44} = \dfrac{37.68}{138.16}$, $\gamma = 15°\ 15'\ 18''$
12	Axial tooth thickness	$s_x = \dfrac{P_x}{2} = \dfrac{9.42}{2} = 4.71$ mm
13	Normal tooth thickness	$s_n = \dfrac{P_x}{2}\cos\gamma = 3.14 \times \dfrac{3}{2} \times \cos\gamma = 4.65$ mm

【Skills Training】

Turning Single Crank Multi-line Worms and Shaft

Processing the workpiece of single crank multi-line worms and shaft as the technique parameters shown in Figure 1.11.

I. Operation preparation

The preparation for processing single crank multi-line worms and shaft is shown in Table 1.13.

Table 1.13　The preparation for turning single crank multi-line worms and shaft

No.	Name		Preparation
1	Material		Round steel $\phi 55$ mm $\times 173$ mm
2	Device		CA6140 turning lathe
3		Cutting tools	$90°$ turning cutter, $45°$ elbow turning cutter, $40°$ thread cutter, A3.15/8mm center drill, grooving cutter
4	Processing tools	Measuring tools	Vernier caliper 0.02 mm / (0-200 mm), micrometer 0.01 mm / (25-50 mm, 50-75 mm), tool setting model, tooth thickness caliper m_x / (1-16 mm), steel ruler, tooth profile template, 10 mm eccentric gasket, magnetic dial indicator 0.01 mm / (0-10 mm)
5		Others	Drill chuck, movable wrench, fixed top, movable top

II . Operation procedure

Steps to process the single crank multi-line worms and shaft is shown in Table 1.14.

Table 1.14　Operation steps of turning single crank multi-line worms and shaft

No.	Operation procedure	Operation diagram
Step 1	The outer circle is clamped tightly with 120 mm stretching out 1) Drill the center hole, hold the workpiece with the top; 2) Rough and fine turning the left end $\phi 35_{-0.025}^{0}$ mm and with a length of 20 mm; 3) Rough and fine turning the right end face $\phi 35_{-0.025}^{0}$ mm and with a length of 20 mm and chamfering C1 mm; 4) Rough and fine turning the diameter of the addendum circle $\phi 50_{-0.03}^{0}$ mm with chamfering $20°$; 5) Rough and fine turning the worms and shaft to size as Figure 1.11 with the tooth top angles being blunted	

No.	Operation procedure	Operation diagram
Step 2	Turn the workpiece around and clamp the copper gasket tightly, turning the end face at the length of 168 mm	
	1) Rough and fine turning the outer circle $\phi35^{0}_{-0.025}$ mm and with a length of 20 mm and chamfering C1mm; 2) Turning the out circle of ϕ50mm	
Step 3	The workpiece is clamped with four-jaw clamping, use the gasket to align, then clamp the workpiece with the eccentricity of 10 mm.	
	1) Drill the center hole on the end face and hold the workpiece with the top; 2) Rough and fine turning the crank $\phi25^{0}_{-0.021}$ mm and with a length of $20^{+0.05}_{0}$ mm, and with a side width of 8 mm, round off the sharp corners	gasket

III. Workpiece quality test

As shown in Figure 1.11, the single crank multi-line worms and shaft should be tested according to its technical requirements.

1. Size and surface roughness

The addendum circle diameter $\phi50^{0}_{-0.03}$ mm shall be tested according to the dimensional tolerance, it is unqualified if out of tolerance.

The pitch circle diameter is $\phi44$ mm，its unspecified tolerance shall be tested according to GB/T1804—2000.

The tooth root circle diameter is $\phi36.8$ mm，its unspecified tolerance shall be tested according to GB/T1804—2000.

The normal tooth thickness $\phi4.55^{-0.025}_{-0.078}$ mm shall be tested according to the dimensional tolerance，it is unqualified if out of tolerance.

The roughness is Ra1.6 μm, unqualified if degradation.

2. Outer diameter and roughness

The outer diameter $\phi 35^0_{-0.025}$ mm shall be tested according to the dimensional tolerance, and it is unqualified if out of tolerance.

The crank diameter $\phi 25^0_{-0.021}$ mm shall be tested according to the dimensional tolerance, and it is unqualified if out of tolerance.

The roughness is $Ra1.6$ μm, and it is unqualified if degradation.

3. Geometric tolerances and eccentricity

The parallelism of the crank shaft relative to the reference shaft is 0.03 mm and it is unqualified if out of tolerance.

The eccentricity of 10 ± 0.04 mm shall be tested according to the dimensional tolerance, and it is unqualified if out of tolerance.

4. Others

The length dimensions are 20 mm, 8 mm, 8 mm, 75 mm, and 20 mm, their unspecified tolerance shall be tested according to GB/T1804—2000.

The roughness of remaining parts is $Ra3.2$ mm, and unqualified if degradation.

IV. Notes

(1) The workpiece must be clamped tightly when turning.

(2) When turning the eccentric shaft, a center hole should be drilled on the end face of the workpiece to prevent the positioning of the workpiece from changing.

(3) When turning the eccentric shaft, it is of necessity to find the parallelism of the positive eccentric shaft and the reference shaft, and the error must be limited within 0.01mm.

Module Two Machining Eccentric Parts and

Multi-throw Crankshaft

Task 1 Machining Double Eccentric Parts

Subtask 1 Machining Double Eccentric Sleeve

【Knowledge and Skills Objectives】

(1) Master the method of scribing double eccentric sleeve.

(2) Master the methods of clamping and machining double eccentric sleeve.

(3) Master the method of measuring the eccentricity of the double eccentric sleeve.

【Related Knowledge】

I. Diagram of double eccentric sleeve

The diagram of double eccentric sleeve is shown in Figure 2.1.

(1) The eccentric sleeve has two symmetrical eccentric holes in the $180°$ direction, and the eccentricity is (12 ± 0.02) mm.

(2) The parallelism tolerance between the $2 \times \phi 18^{+0.021}_{0}$ mm eccentric circle center line and datum axis center line of the outer circle is $\phi 0.02$ mm.

(3) The parallelism tolerance of the right end to the left end is 0.03 mm.

(4) The outer circle with diameter $\phi 62^{0}_{-0.041}$ mm is the reference circle.

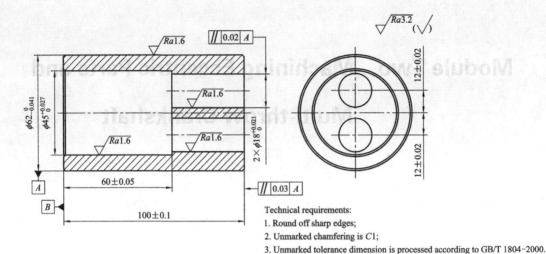

Figure 2.1   Diagram of double eccentric sleeve

Technical requirements:
1. Round off sharp edges;
2. Unmarked chamfering is $C1$;
3. Unmarked tolerance dimension is processed according to GB/T 1804–2000.

II. The steps to process double eccentric sleeve

1. Clamping the workpiece

Because the workpiece is processed in single-piece production, a four-jaw single-action chuck is used to clamp the workpiece for processing. Rough turning the end face and the outer circle first, then fine turning the inner circle and outer circle, and finally processing two eccentric holes.

2. Processing heat treatment

The heat treatment for tempering should generally be arranged after rough turning and before finishing turning. Because it is processed in single-piece production, the heat treatment can also be arranged before the rough turning to simplify the process.

3. Aligning the eccentricity

1) Measuring the eccentricity

It is very difficult to guarantee the machining accuracy of the eccentricity by scribing and aligning processes. The accuracy requirements must be guaranteed by measuring with the gauge block. The method shown in Figure 2.2(a) can be used. When clamping the workpiece for aligning, the gauge block is placed on one end of the workpiece, and the size of the gauge block is equal to twice the eccentricity. A dial indicator can be used to measure the eccentricity and adjust it to make the positions of a and b be the highest and the lowest.

2) Calibrating and scribing

In order to ensure that the centers of the two eccentric holes are symmetrically distributed at $180°$, after one eccentric hole is turned, find the thinnest point between the outer circle $\phi62_{-0.041}^{0}$ mm and the eccentric hole, draw a straight line from the thinnest point and the centre of the hole, then the position of another symmetrical hole is on this straight line. Scribe and align this straight line with a $90°$ angle ruler and draw a cross line as shown in Figure 2.2(b).

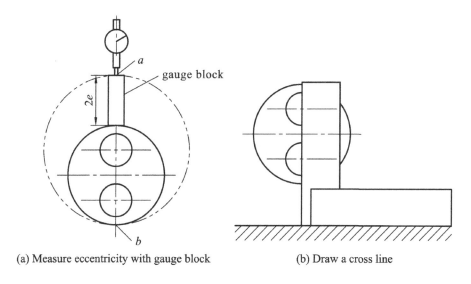

(a) Measure eccentricity with gauge block   (b) Draw a cross line

Figure 2.2 Scribing and aligning the eccentricity

III. Scribing and Aligning when the double eccentric part loaded on the four-claw single-action chuck

The method of aligning the cross lines and the lateral generatrix of the part loaded on the four-jaw single-action chuck is shown in Figure 2.3.

1. Scribing and aligning the cross-line and the lateral generatrix

First, mark the cross lines and the lateral generatrix on both ends of the workpiece. When aligning the cross line and the lateral generatrix, the scribing plate should be placed on the small plate (the verticality of the end face of the processed workpiece should be modified with a width of a $90°$ angle ruler).

(1) Release the jaws according to the size of the workpiece. The positions of four jaws can be preliminarily determined according to the arcs of the concentric circles on the end face of the chuck.

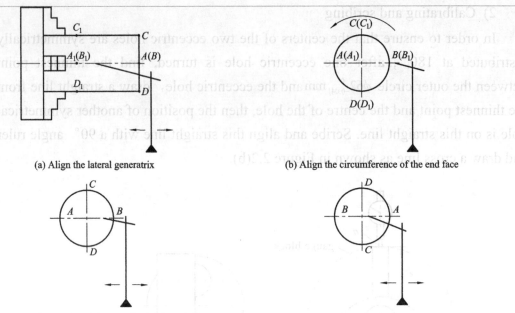

(a) Align the lateral generatrix   (b) Align the circumference of the end face

(c) Align the AB line of the cross line of the end face   (d) Align the needle tip through the AB line after rotating the workpiece 180°

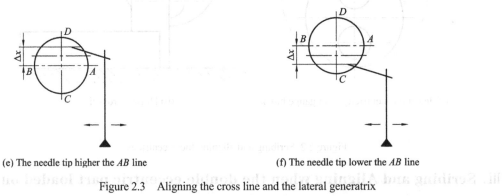

(e) The needle tip higher the AB line   (f) The needle tip lower the AB line

Figure 2.3   Aligning the cross line and the lateral generatrix

(2) Align the lateral generatrix of AA, or BB_1, and CC_1 or DD_1 with the needle tip of the scribing plate, and move the scribing plate axially to correct the horizontal position of the workpiece (see Figure 2.3(a)).

(3) Align the tip of the plate with the circumferential line of the end face of the workpiece, rotate the chuck, and initially align the coaxial line between the axis and the main axis line(see Figure 2.3(b)).

(4) When aligning the cross line of the end face of the part, it is unknown whether the needle tip is at the same height as the main axis line. When aligning, the height of the needle tip and the position of the workpiece must be adjusted at the same time. First, pass the needle tip through the end face AB line (see Figure 2.3(c)), and then turn the workpiece through 180° , the needle tip will pass through the AB

line again. At this time, three situations may occur as follows.

① The needle tip still passes the *AB* line, which means that the needle tip is at the same height as the main axis line, and the workpiece *AB* line passes through the axis, as shown in Figure 2.3(d).

② The needle tip is higher than the *AB* line and the distance is Δx, as shown in Figure 2.3(e). At this time, the needle should be adjusted downward by $\Delta x/2$ to the axis initial height, and the *AB* line of the workpiece should be adjusted upward by $\Delta x/2$ to the axis initial position.

③ The needle tip is lower than the *AB* line and the distance is Δx, as shown in Figure 2.3(f), the needle tip should be adjusted upward by $\Delta x/2$, and the *AB* line of the workpiece should be adjusted downward by $\Delta x/2$. In this way, the workpiece is repeatedly turned 180° for alignment until the situation shown in Figure 2.3(c), Figure 2.3(d) appears. When the scribing height is adjusted, it is much easier to align the *CD* line with the needle tip. After aligning the *CD* line, the positive lateral generatrix can be aligned, so repeatedly, finally the aligned axis can coincide with the main axis line.

2. Scribing and aligning the Tian-shaped frame line of the double eccentric sleeve

The diagram of double eccentric sleeveis is in Figure 2.1, this workpiece is clamped in the V-shaped groove above the square box, and the eccentric Tian-shaped frame line is scribed on the square box, as shown in Figure 2.4(a).

Move the vernier height ruler down from the highest point of the workpiece by a distance of $D/2$, draw the center line on the two end faces and the lateral generatrix on the outer circle, then draw the eccentric center line according to the eccentricity, finally draw the frame line of the round hole according to the shape of the hole (the Tian-shaped frame line). After drawing the above lies, turn over the square box for 90° , and draw the center cross line on the two end faces, and draw the lateral generatrix on the outer circle. At this time, the center cross line of the end face, the processing frame line of the end surface and the lateral generatrix on the outer circle are formed.

The frame line of the circular hole (see Figure 2.4(b)) is good for locating the top, the bottom, the left and the right of the hole, which has a high reference value when aligning and processing the part. The Tian-shaped frame line of the circular hole is helpful for the alignment of the center cross line. It is more accurate to make the alignment by pointing the scribe needle on the frame line and rotating the workpiece. The frame line constrains the position of the hole during processing, which makes the position of the hole more accurate. After the lines have been drawn, a punching hole can be drawn in the center of the cross line, and a round hole line can be drawn by a compass. The round hole line can be used as the positioning reference during processing.

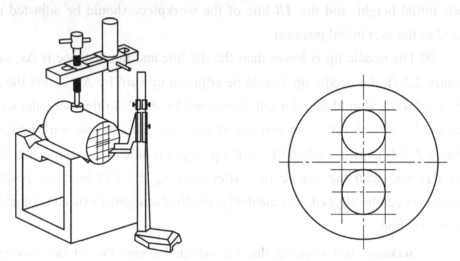

(a) Machine to draw the eccentric Tian-shaped frame line on the square box   (b) The hole in the eccentric Tian-shaped frame line

Figure 2.4   Drawing the eccentric Tian-shaped frame line

In the above example, the outer circle of the eccentric sleeve is the positioning reference, and the eccentric holes on both sides are in reverse offset by $180°$. During processing, a four-jaw single-action chuck is used to clamp the workpiece, and the scribing plate is used to align the center cross line, eccentric center cross line on the two-end and the lateral generatrix. A dial gauge should be used to press the outer circle to correct the offset of the inner hole, or correct and check the accuracy of the size by the Tian-shaped frame line.

IV. Detection of the double-eccentric holes

The double-eccentric holes are the key detection parts, and the detection parameters (eccentricity, hole spacing size and parallelism, and their detection diagrams are as Figure 2.5). Insert two inspection rods into the eccentric holes $\phi 18^{+0.021}_{0}$ mm, then place the part on a V-shaped iron and use a dial gauge to check the eccentricity dimensions of the double-eccentric holes based on the outer diameter, as shown in Figure 2.5(a). Two inspection rods with an outer micrometer can be used to check the hole spacing size, as shown in Figure 2.5(b). Move the dial indicator axially to check the parallelism $\phi 0.02$ mm of the eccentric hole to the axis of the part, as shown in Figure 2.5(c). The above methods can be used simultaneously to measure the eccentricity, parallelism and distance (hole spacing size) between two holes $\phi 18_{\ 0}^{+0.021}$ mm.

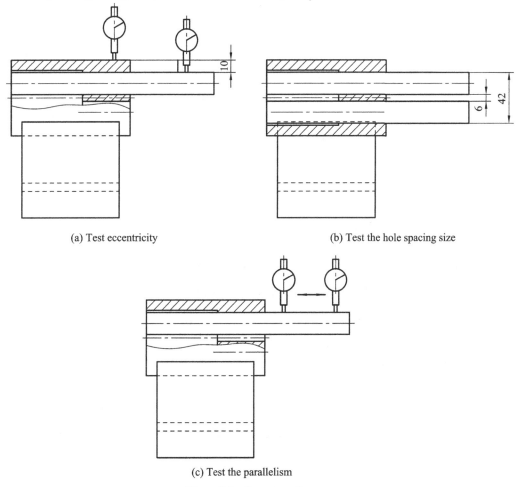

(a) Test eccentricity   (b) Test the hole spacing size

(c) Test the parallelism

Figure 2.5   Detection diagrams

【Skills Training】

Turning Double Eccentric Sleeve

Processing the double eccentric sleeve, whose parameters are shown as in Figure 2.1.

I. Operation preparation

The preparation for processing double eccentric sleeve is shown in Table 2.1.

Table 2.1 The preparation for machining double eccentric sleeve

No.	Name		Preparation
1	Material		Round steel $\phi67$mm$\times$120mm
2	Device		CA6140 turning lathe, a four-claw single-action chuck
3	Processing	Cutting tools	90° turning cutter, cut-off tool, 45° elbow turning cutter, inner hole rough turning cutter, inner hole fine turning cutter, drill $\phi16$ mm, drill $\phi42$ mm
4		Measuring tools	Universal protractor 2'(0° -320°), micrometer 0.01 mm / (0-25 mm, 25-50 mm, 50-75 mm), magnetic seat dial indicator 0.01 mm / (0-30 mm), inner hole dial indicator 0.01 mm / (0-30 mm), vernier height ruler 0.02 mm / (0-300 mm)
5		Others	Square box, slotted screwdriver, adjustable wrench, center, drill fixture, scribing plate

II. Operation procedure

The operation steps of processing double eccentric sleeve is shown in Table 2.2.

Table 2.2 Operation steps of processing double eccentric sleeve

No.	Operation procedure	Operation diagram
Step 1	Tempering treatment HBW220-250; The workpiece is clamped with 103 mm stretching out 1) Turning the end face; 2) Drill a hole $\phi42$ mm with a length of 59 mm; 3) Rough and finish turning outer circle $\phi62_{-0.041}^{0}$ mm with a length of 101mm; 4) Chamfer $C1$ mm	
Step 2	1) Rough turning inner hole $\phi44$mm with a length of 59.5mm; 2) Finish turning inner hole $\phi45_{0}^{+0.027}$ mm with a length of (60 ± 0.05) mm; 3) Cut the parts to ensure a length of 101 mm; 4) Turn the workpiece around and turning the end face to ensure the total length of (100 ± 0.10) mm	

Continued Table

No.	Operation procedure	Operation diagram
Step 3	Scribe lines on the workpiece above the square box	
	Draw the Tian-shaped frame line	
Step 4	Clamp, align and turning the workpiece	
	1) Lightly clamp the workpiece with 50mm stretching out by the copper gasket soft claw of the four-jaw single-action chuck and find the eccentricity (12 ± 0.02) mm and parallelism; 2) Drill the hole of $\phi 16$mm; 3) Rough turning the inner hole of $\phi 17.9$ mm; 4) Finish turning the inner hole of $\phi 18^{+0.021}_{0}$ mm; 5) Apply the same method to turn another eccentric hole	

Ⅲ. Workpiece quality test

The double eccentric sleeve should be tested according to its technical requirements shown in Figure 2.1.

1. Outer diameter and roughness

The external diameter $\phi 62^{0}_{-0.041}$ mm was tested according to the dimensional tolerance, and it is unqualified if out of tolerance.

Roughness is $Ra1.6$ μm, or unqualified if downgrading.

2. Inner diameters, eccentricity, and roughness

The inner diameters $\phi 45^{+0.027}_{0}$ mm and $\phi 18^{+0.021}_{0}$ mm (2 places) are all tested according to dimensional tolerance, and they are unqualified if out of tolerance.

The eccentricity is 12 ± 0.02 mm shall be tested according to the dimensional tolerance, and it is unqualified if out of tolerance.

Roughness is $Ra1.6$ μm, or unqualified if downgrading.

3. Length dimensions

The length dimensions of 60 ± 0.05 mm and 100 ± 0.1 mm are tested according to the dimensional tolerance, and they are unqualified if out of tolerance.

The length dimension is 40 mm, its unspecified tolerance shall be tested according to GB/T1804—2000.

4. Geometric tolerance

The parallelism of 0.03 mm is tested according to the geometric tolerance, and it is unqualified if out of tolerance.

5. Others

The remaining chamfers are $C1$ mm (2 places), their unspecified tolerance shall be tested according to GB/T1804—2000.

Other roughness is $Ra3.2$ μm or unqualified if downgrading.

IV. Notes

(1) When turning the eccentric hole, according to the scribing reference line, the aforementioned gauge block measurement method is used to correct the eccentricity of the workpiece.

(2) The parts should be clamped firmly to avoid clamping injuries.

(3) A lower cutting speed should be adopted when rough turning, and the rotation speed of the spindle should be increased when finish turning.

Subtask 2   Turning Coaxial Double Eccentric Shaft and Reverse Double Eccentric Sleeve

【Knowledge and Skills Objectives】

(1) Master the processing method of coaxial double eccentric shaft and reverse double eccentric sleeve.

(2) Master the eccentricity measuring method of coaxial double eccentric shaft and reverse double eccentric sleeve.

【Related Knowledge】

Ⅰ. Diagrams of the coaxial double eccentric shaft and the reverse double eccentric sleeve

The diagram of coaxial double eccentric shaft is shown as in Figure 2.6，the diagram of reverse double eccentric sleeve(bushing) is shown as in Figure 2.7.

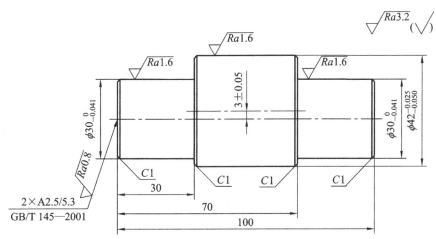

Technical requirements:

Unmarked tolerance dimension is processed according to GB/T 1804—2000.

Figure 2.6   The diagram of coaxial double eccentric shaft

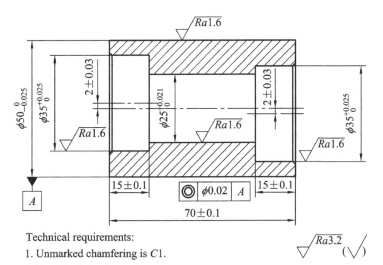

Technical requirements:

1. Unmarked chamfering is C1.

2. Round off the sharp edges.

3. Unmarked tolerance dimension is processed according to GB/T 1804—2000.

Figure 2.7   The diagram of reverse double eccentric sleeve

1. The coaxial double eccentric shaft

(1) The reference shaft is an intermediate shaft, and both sides have eccentric shafts in the same direction.

(2) Drill center holes on both sides.

2. The reverse double eccentric sleeve

(1) The reference shaft is an outer circular axis, and the center hole has coaxial requirement relative to the reference shaft.

(2) Both sides have reverse eccentric holes.

II. Clamping methods for double eccentric parts

1. Method of clamping with two supporting tops

Method of clamping with two supporting tops use two tops to support the eccentric center holes on both ends of the workpiece to get an eccentric shaft. This clamping method requires to align the eccentric center holes on both ends accurately, and it is suitable for turning long eccentric shaft, as shown in Figure 2.8. In this method, four center holes must be drilled on the two end faces of the workpiece according to the requirements of the eccentric distance e, and then the center holes can be turned with supporting tops. Diagram of the long eccentric shaft is shown in Figure 2.9.

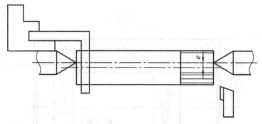

Figure 2.8　Turning the long eccentric shaft with method of two supporting tops

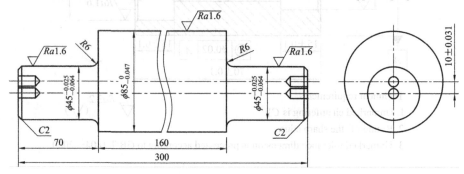

Figure 2.9　The diagram of the long eccentric shaft

The procedures to turn the eccentric shaft are as follows: first rough-turning the outer circle of ϕ86 mm with a length of 301 mm, cut the workpiece and turn the end face to ensure a total length of 300 mm; then drill four center holes on the two end faces; finally, top against the reference center hole, and finish turning the outer circle $\phi85_{-0.047}^{0}$ mm, top against the eccentric center hole, the outer circle is roughed and refined to the size $\phi45_{-0.064}^{-0.025}$ mm.

2. Method of clamping the workpiece with a combination of the four-jaw single-action chuck and the three-jaw self-centering chuck

When turning a large number of eccentric workpieces, it will take a lot of time to align the workpieces. At this time, the clamping method should be changed, that is, the three-jaw self-centering chuck is installed on the four-jaw single-action chuck, the four-jaw single-action chuck is used for adjusting eccentricity, and put the workpiece directly on the three-jaw self-centering chuck for calibration, the diagram of clamping the part with the two tools combination is shown as in Figure 2.10.

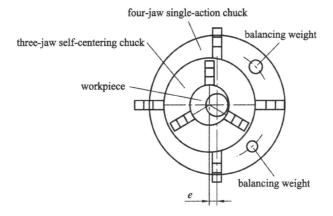

Figure 2.10   The diagram of clamping with the combination of three-jaw
self-centering chuck and the four-jaw single-action chuck

When processing, it just needs to align the first workpiece, then it is unnecessary to align each workpiece, which can save a lot of auxiliary time. When using this method, the four-jaw single-action chuck and the three-jaw self-centering chuck are required to have higher accuracy, especially for the three-jaw self-centering chuck. When in calibrating the first workpiece, align the parallelism of the end face of the workpiece under the premise of aligning the outer circle. If the claw feet of the four-jaw single-action chuck are not in contact, a thin gasket can be used to fill in to prevent the workpiece from position shift or deformation during clamping and turning. This method is only suitable for turning eccentric workpieces with small size, short length, small eccentricity, and low precision requirements. If the eccentricity of the workpiece is large, the clamping force of the four-jaw single-action chuck would be unevenly distributed on the outer circle of the three-jaw self-centering chuck, and the clamping force would be reduced, a great centrifugal force would make the clamping unreliable.

This method requires the use of balancing weights to avoid vibration and ensure safety.

3. Method of clamping the eccentric workpiece on the faceplate

The inner holes of a small number of eccentric workpieces can be clamped and turned on the faceplate. The diagram of clamping the workpiece on the faceplate is shown in Figure 2.11. Steps to turning workpiece on the faceplate are as follows: first turning the outer circle of the eccentric workpiece (to ensure that it is within a certain tolerance); then clamp the workpiece on the faceplate. Pre-align the circumferential line drawn with the eccentricity crossing the center of the circle, press the two pressure plates firmly, install locating blocks on the outer circle of the workpiece far from the hole, and ensure that the locating blocks are distributed at a right angle of $90°$. This method does not require to use special fixture, the processing accuracy is high, and the clamping is firm and reliable.

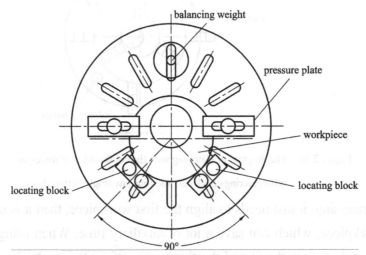

Figure 2.11   Diagram of clamping the workpiece on the faceplate

4. Method of clamping mass production of eccentric workpieces with special fixture

For turning mass production of eccentric workpieces, special fixture is generally used. This method can shorten the alignment time, not only ensure workpiece quality, but also improve production efficiency. A special fixture—eccentric bushing, its diagram is shown as in Figure 2.12. It is used to clamp and turn the eccentric shaft.

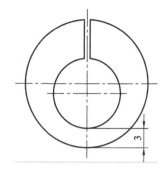

The method of turning the eccentric shaft clamped by the eccentric bushing is as follows: The eccentric bushing is processed in accordance with the requirements of the workpiece's eccentricity. The eccentricity of the eccentric bushing must be strictly guaranteed to be at the middle value of the eccentricity tolerance of the eccentric shaft, so as to decrease the risk of producing waste products caused by eccentric bushing's installation error. The surface roughness of the eccentric hole should reach $Ra1.6$ μm or more, the eccentric bushing has a gap of 1–2 mm along the axial direction, and the smallest wall thickness is guaranteed to be about 3 mm.

Figure 2.12   Diagram of a special fixture —eccentric bushing

The eccentric wheel can be processed according to the above principle. The eccentric wheel clamped with eccentric shaft is as shown in Figure 2.13. The eccentric wheel is turned in advance, and then the eccentric wheel is sleeved on the eccentric shaft to fine turning the outer circle of the eccentric wheel. This method uses the eccentric shaft as the mandrel. When turning, attention must be paid to ensure sufficient tightening force, and the gap between the eccentric shaft and the eccentric hole should not be too large.

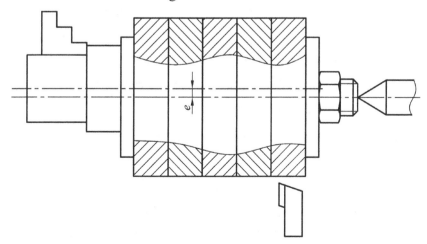

Figure 2.13   Clamping the eccentric wheel with eccentric shaft

Above special fixture used for clamping eccentric holes benchmarked by the outer circle of eccentric workpiece is shown in Figure 2.14. This fixture is simple and accurate, and can save the time of aligning the workpiece.

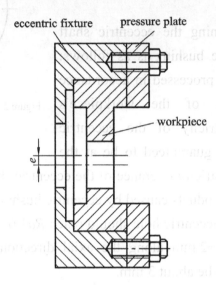

Figure 2.14   The special fixture used for clamping eccentric holes

III. Measurement methods for eccentric workpieces

1. Vernier caliper measuring method

Vernier caliper measuring method is the simplest measurement method, suitable for measuring eccentric workpieces that do not require high measurement accuracy. When in use, the thickest part (maximum size) and the thinnest part (minimum size) of the eccentric hole wall of the workpiece should be measured, so as to accurately measure the value of the center distance. Half of the difference between the measured maximum size and the measured minimum size is the eccentricity.

2. Dial indicator measuring method

Dial indicator measuring method is suitable for eccentric workpieces with high accuracy requirements and small eccentricity. The diagram used for measuring the eccentric wheel with dial indicator is shown as in Figure 2.15. In this method, it takes the hole as the reference plane, makes a reference axis on the three-jaw self-centering chuck and puts the surface of the wheel tightly on the claw surface. The contact of the dial indicator is placed on the outer circle of the eccentric wheel, and then slowly rotate the eccentric wheel, the difference between the maximum value and the minimum value of the dial indicator should be twice the eccentricity, otherwise, the eccentricity does not meet the requirements.

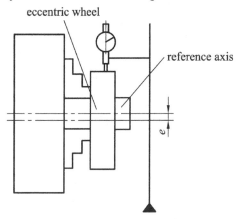

Figure 2.15   The diagram of measuring the eccentric wheel with a dial indicator

This method can also be used to measure the reverse eccentric sleeve according to the parameters of reverse double eccentric sleeve in Figure 2.7. Place the reverse double eccentric sleeve on the reference axis, press the lever percentage gauge at the eccentric hole position, and slowly rotate the double eccentric sleeve to measure the eccentricity, the diagram of measuring the reverse eccentric sleeve with a dial indicator is shown as in Figure 2.16.

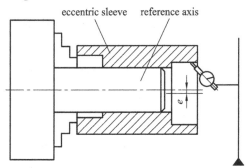

Figure 2.16   The diagram of measuring the reverse eccentric sleeve with a dial indicator

3. Measurement method of combination of a dial indicator and V-shaped frames

The diagram of measuring the eccentric shaft with a dial indicator and V-shaped frames is shown as in Figure 2.17. First, place the workpiece on a V-shaped frame or two equal-height V-shaped irons and the dial indicator contact on the outer circle of the eccentric part. Then, rotate the eccentric shaft, and the reading on the dial indicator should be equal to twice the value of the eccentricity of the workpiece.

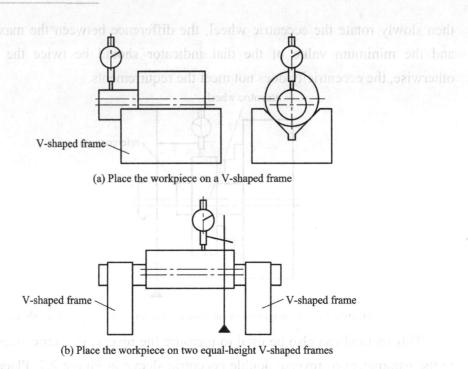

(a) Place the workpiece on a V-shaped frame

(b) Place the workpiece on two equal-height V-shaped frames

Figure 2.17   The diagram of measuring the eccentric shaft with a dial

indicator and V-shaped frames

4. Measuring method of combination of dial indicator and middle slide plate of lathe

As shown in Figure 2.18, the eccentric part with a large eccentricity can be measured on a middle slide plate of lathe with a dial indicator, and the scale of the middle slide plate of lathe can be used to compensate the measuring range of the dial indicator, which can obtain more accurate measurement results. When measuring, first contact the dial indicator with the highest point on the eccentric outer circle of the workpiece, record the reading of the dial indicator and the scale value of the middle slide plate, and then turn the workpiece through $180°$ and screw it into the middle slide plate to make the movement distance of the middle sliding plate equal to twice the eccentricity, and thus, the deviation between the dial gauge reading and the original reading is the deviation of the eccentricity. When using this method to measure a long eccentric workpiece, attention should be paid to the parallelism between the main axis and the eccentric axis of the workpiece. When measuring, move the contacts of the dial indicator to the two ends of the workpiece, observe the dial indicator reading. If the error value is within the allowable range, the parallelism of the workpiece is correct, otherwise, it does not meet the requirements.

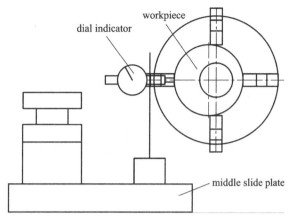

Figure 2.18   The diagram of measuring the workpiece with a large eccentricity on

the middle slide plate of the lathe with a dial indicator

This method can be applied not only to measuring the workpiece with a large eccentricity distance, but also measuring the workpiece with a small eccentricity distance. In the measurement, as long as the measuring range of the dial indicator is greater than twice the eccentricity distance, rotate the workpiece to measure its eccentricity, which is widely used in production (the marking line of the middle slide plate of lathe should be adjusted accurately).

【Skills Training】

Training 1

Turning Coaxial Double Eccentric Shaft

When machining coaxial double eccentric shaft, its parameters are shown in Figure 2.6, after the turning, the eccentricity of the two ends of the workpiece should be the same, and the offsets of center holes at both ends in the circumferential direction should be exactly the same. In this training task, an eccentric fixture is used to clamp the workpiece, drill the eccentric center holes at both ends, and then two tops are used to locate the workpiece for turning. The eccentric fixture is a self-made eccentric sleeve, we can insert the workpiece into the eccentric fixture and fix it with screws, the diagram of an eccentric fixture is shown as in Figure 2.19.

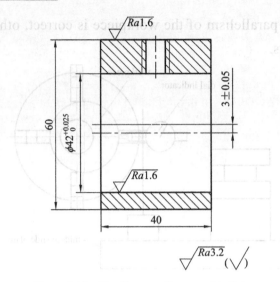

Figure 2.19   The diagram of an eccentric fixture

This method can ensure that the axes of the four eccentric central holes at both ends of the part are in the same vertical plane, and the central holes at both ends are not offset in the circumferential direction, it is a simple and convenient method.

I . Operation preparation

The preparation for turning the coaxial double eccentric shaft is shown in Table 2.3.

Table 2.3   The preparation for turning the coaxial double eccentric shaft

No.	Name		Preparation
1	Material		45 steel, ϕ47mm×130mm
2	Device		CA6140 turning lathe
3	Processing tools	Cutting tools	90° rough turning cutter, 90° finish turning cutter, 45° elbow turning cutter, center drill A2.5/6.3 mm
4		Measuring tools	Vernier caliper 0.02 mm / (0−150 mm), micrometer 0.01 mm / (25−50 mm), magnetic seat dial indicator 0.01 mm / (0−30 mm)
5		Others	Self-made eccentric sleeve, the copper skin, screwdriver for slotted head screws, adjustable wrench, top, drilling fixture, etc.

II . Operation procedure

The operation steps for turning the coaxial double eccentric shaft are shown in Table 2.4.

Table 2.4   Operation steps for turning the coaxial double eccentric shaft

No.	Operation procedure	Operation diagram
Step 1	Install the workpiece on the three-jaw self-centering chuck, and clamp it with 110 mm stretching out 1) Turning the end face; 2) Rough and fine turning the outer circle $\phi 42_{-0.050}^{-0.025}$mm ; 3) Cut the workpiece to ensure the length of 102 mm 4) Turn the workpiece around and turning the end face to ensure the total length of 100 mm	
Step 2	Place the eccentric sleeve in the jaws Clamp the workpiece in the eccentric sleeve and drill the eccentric center holes at both ends	
Step 3	Turning the workpiece supported with two tops 1) Place the workpiece between the two tops, and turning the eccentric shaft $\phi 30_{-0.041}^{0}$mm $\times$ 30mm on both sides; 2) Chamfer $C1$ mm	

III. Workpiece quality test

The coaxial double eccentric shaft should be tested according to the technical requirements shown in Figure 2.6.

1. The outer diameter and roughness

The external diameter $\phi 42_{-0.050}^{-0.025}$mm and $\phi 30_{-0.041}^{0}$mm (2 places) shall be all tested according to dimensional tolerance, and it is unqualified if out of tolerance.

Roughness is Ra1.6 μm or unqualified if downgrading.

2. Eccentricity

The eccentricity of 3 ± 0.05 mm is tested according to the dimensional tolerance, and it is unqualified if out of tolerance.

3. Length dimensions

The length dimensions of 100 mm, 70 mm, and 30 mm are all tested, their unmarked tolerance is tested according to GB/T1804—2000.

4. Others

Roughness in other positions is $Ra1.6$ μm, or unqualified if downgrading.

IV. Notes

The eccentric fixture is a self-made eccentric sleeve that needs to be prepared in advance. Before turning, insert the parts into the eccentric fixture and fix them with screws. When turn the parts around, the parts in the eccentric fixture should be clamped tightly.

Training 2

Turning Reverse Double Eccentric Bushing

Processing the reverse double eccentric bushing(sleeve) according to its technique parameters shown as in Figure 2.7.

I. Operation preparation

The preparation of processing the reverse double eccentric bushing is shown in Table 2.5.

Table 2.5　The preparation of processing the reverse double eccentric bushing

No.	Name		Preparation
1	Material		Medium carbon steel, ϕ55 mm×100 mm
2	Device		CA6140 turning lathe
3	Processing tools	Cutting tools	90° rough turning cutter, 90° finish turning cutter, 45° elbow turning cutter, inner hole rough turning cutter, inner hole finish turning cutter, drilling bit of ϕ23 mm and ϕ32 mm
4		Measuring tools	Vernier caliper 0.02 mm/(0−150 mm), micrometer 0.01mm/(25−50 mm), inner hole dial indicator 0.01 mm/(18−35 mm), magnetic seat dial indicator 0.01 mm/(0−30 mm)
5		Others	Copper skin, adjustable wrench, screwdriver, top, drilling fixture, etc.

II. Operation procedure

The operation steps of processing the reverse double eccentric bushing are shown in Table 2.6.

Table 2.6   Operation steps of processing the reverse double eccentric bushing

No.	Operation procedure	Operation diagram
Step 1	Clamp the workpiece with 75 mm stretching out 1) Turning the end face; 2) Drill a hole with the diameter $\phi23$ mm and with the length of 75 mm; 3) Rough and fine turning the outer circle to size $\phi50_{-0.025}^{0}$ mm; 4) Rough and finish turning the inner hole to size $\phi25_{0}^{+0.021}$ mm; 5) Cut the workpiece to ensure the total length of 71 mm	
Step 2	Apply process ink or copper sulfate solution on the end face and outer circle of the workpiece, and scribe the shape of workpiece on the square box Place the workpiece on the square box; draw the center line, the Tian-shaped frame line on the end face and the lateral generatrix on the outer circle	
Step 3	Clamp the workpiece with a four-jaw single-action chuck on the lathe 1) Use the scribing plate to align the cross line on one end face and the lateral generatrix on the outer circle, and use a dial indicator to verify the accurate value of the eccentricity. Use the same method to align the other end face, turning and measuring the workpiece by referring to the Tian-shaped frame line	

Continued Table

No.	Operation procedure	Operation diagram
Step 3	2) Turning an eccentric hole with a diameter $\phi 35^{+0.025}_{0}$ mm and a length of 15mm, $Ra1.6\,\mu m$; 3) Chamfer the inner hole $C1$mm	
Step 4	The other end face is turned as Step 3	

III. Workpiece quality test

The reverse double eccentric sleeve should be tested according to its technical requirements shown in Figure 2.7.

1. Outer diameter and roughness

The outer diameter (diameter of outer circle) $\phi 50^{0}_{-0.025}$ mm is tested according to the dimensional tolerance, and it is unqualified if out of tolerance.

Roughness is $Ra1.6\,\mu m$, or unqualified if downgrading.

2. Inner diameter and roughness

The inner diameter (diameter of inner hole) $\phi 25^{+0.021}_{0}$ mm is tested according to the dimensional tolerance, and it is unqualified if out of the tolerance.

Roughness is $Ra1.6\,\mu m$, or unqualified if downgrading.

3. Geometric tolerance and eccentricity

The coaxiality $\phi 0.02$ mm is tested according to the geometric tolerance, and it is unqualified if out of tolerance.

The eccentricity of 2 ± 0.03mm is tested according to the dimensional tolerance, and it is unqualified if out of the tolerance.

4. Length dimensions

The lengths of 15 ± 0.1mm (2 places) and 70 ± 0.1mm are tested according to the dimensional tolerance, and they are unqualified if out of tolerance.

5. Others

The remaining chamfers are $C1$mm and their unmarked tolerance should be tested according to GB/T 1804—2000.

Roughness is $Ra3.2$ μm, or unqualified if downgrading.

IV. Notes

1. Coaxiality requirements

There are coaxiality requirements with an error of 0.02mm for the diameter of the outer circle $\phi 50_{-0.025}^{0}$ mm and the diameter of the nner hole $\phi 25_{0}^{+0.021}$ mm. To ensure the tolerance, the inner hole and the outer circle should be completed in one processing.

2. Scribing, aligning, processing sequence

The two inner holes are $180°$ reverse symmetrical eccentric holes. After the outer circle is processed, scribing, aligning and processing are performed in sequence.

Task 2   Machining Four-throw Crankshaft

【Knowledge and Skills Objectives】

(1) Master the method of clamping and machining a four-throw crankshaft.

(2) Master the causes and solutions of deformation when machining crankshaft.

(3) Master the method of measuring crankshaft.

【Related Knowledge】

I. Diagram of the four-throw crankshaft

The diagram of four-throw crankshaft is shown as in Figure 2.20.

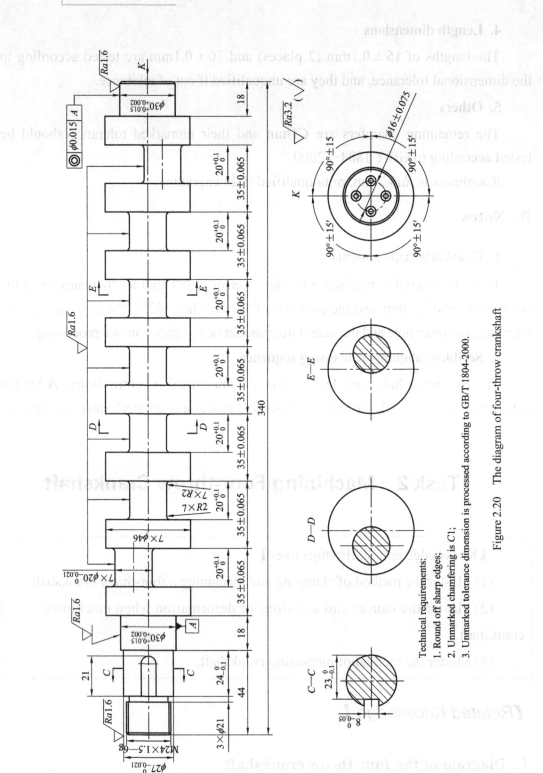

Technical requirements:
1. Round off sharp edges;
2. Unmarked chamfering is C1;
3. Unmarked tolerance dimension is processed according to GB/T 1804-2000.

Figure 2.20　The diagram of four-throw crankshaft

The workpiece is a four-throw crankshaft, each crank with a diameter of $\phi 20_{-0.021}^{0}$ mm and a length of $20_{0}^{+0.1}$ mm, and the main journal with the diameter of $\phi 30_{+0.002}^{+0.015}$ mm and a length of 18 mm long. The diameter of the keyway shaft at the left end of the main journal is $\phi 27_{-0.021}^{0}$ mm, with the width of the keyway $8_{-0.05}^{0}$ mm and the length of the keyway $20_{-0.1}^{0}$ mm. The left end thread is M24 × 1.5 fine thread and the length of the thread shaft and the keyway shaft is 44 mm, and the diameter of the eccentric center circle of the crankshaft is $\phi (16 \pm 0.075)$ mm.

II. The method of clamping crankshaft

1. Clamping crankshaft with one clamp and one top

When turning or grinding the crankshaft, the main work is to solve the problem of workpiece clamping, that is to say, how to align the axis of the crank journal, coincide it with the axis of rotation of the spindle of the lathe or grinder.

When the diameter of the crankshaft is relatively thick and the eccentricity is small, this method is to clamp the chuck on the faceplate so that the distance between the chuck axis and the main shaft axis is equal to the eccentric distance of the crankshaft journal. Diagram of clamping the crankshaft with one clamp and one top is shown in Figure 2.21, first drill a central hole A on both ends, rough turning the main journal, the diameter of the connecting plate and its lengthened part by one clamp and one top method; then drill the center holes $B, C, D,$ and E of each crank journal on the end face of the connecting plate. Clamp the main journal d with a chuck and the eccentric center hole supported by the top to ensure that the indexing of the crank journal and its parallelism between the crankshaft journal and the main journal.

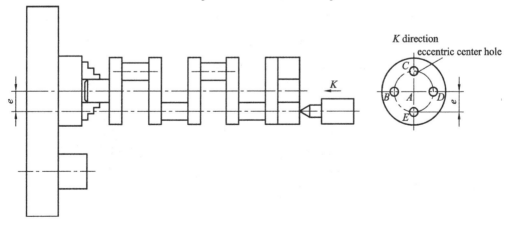

Figure 2.21   Diagram of clamping the crankshaft with one clamp and one top

2. Clamping crankshaft with two tops

Diagram of clamping the crankshaft with two tops is shown in Figure 2.22, first, drill the central hole A of the main journal and the eccentric center holes B, C, D, and E of the crank journal on both ends of the crankshaft; then use each center hole as positioning datum, install the crankshaft on the two tops of the machining tool, respectively process the crank and the outer circle; finally process the main shaft journal, and turning the eccentric center holes B, C, D, E on both ends. This method is suitable for crankshaft with small size or for shafts with small eccentricity and it is conveniently used for positioning and clamping. Generally, the workpiece is directly processed with round bar materials. The position accuracy between the crank journal and the main shaft journal is guaranteed by that of the eccentric center holes at both ends.

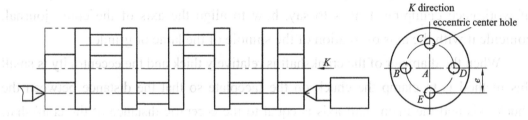

Figure 2.22   Diagram of clamping the crankshaft with two tops

3. Clamping crankshaft with eccentric splints

The crankshaft clamped with eccentric splints is shown as in Figure 2.23. First, install a pair of eccentric splints on the main journals (the diameter is the processing size with a margin) at both ends of the processed crankshaft, and use tools such as V-shaped blocks on the flat plate to align the workpiece. After aligning, tighten the screws on the eccentric splints; and then machining the crank journal through the eccentric center hole supported by the tops at both ends of the eccentric splint. This method is suitable for crankshafts that have a large eccentricity and whose eccentric center holes on the end face cannot be drilled easily.

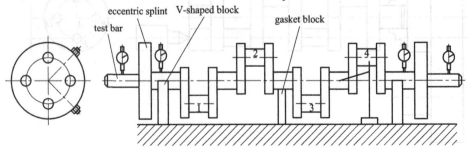

Figure 2.23   Diagram of clamping the crankshaft with eccentric splints

4. Clamping crankshaft with one eccentric chuck

The eccentric chuck is a commonly used special fixture for processing eccentric workpieces, which is suitable for processing eccentric workpieces with high eccentric precision. The diagram of eccentric chuck is shown as in Figure 2.24, the eccentric chuck is divided into two layers, the faceplate 2 is installed on the main spindle of the lathe with screws, and the eccentric body 3 is installed on the faceplate dovetail groove. A three-jaw self-centering chuck 5 is installed on the eccentric body 3. The lead screw 1 is used to adjust the eccentricity of the chuck, and the eccentricity can be measured between the two measuring heads 6 and 7. When the eccentricity is zero, the measuring heads 6 and 7 would just collide. When the screw 1 is rotated, the measuring head 7 gradually leaves the measuring head 6, and the deviation is the eccentricity. The deviation between the two measuring heads can be measured by a dial indicator and gauge blocks. When the eccentricity is adjusted, use

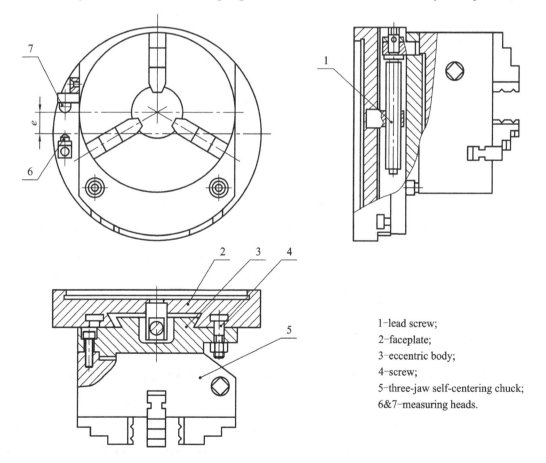

1–lead screw;
2–faceplate;
3–eccentric body;
4–screw;
5–three-jaw self-centering chuck;
6&7–measuring heads.

Figure 2.24   The diagram of eccentric chuck

screw 4 (four screws)to tighten and clamp the workpiece on the three-jaw self-centering chuck, and then turning can be processed. Since the eccentricity of the eccentric chuck is measured by a dial indicator and gauge blocks, a high eccentricity accuracy can be obtained. The part clamped with eccentric chuck compared to that clamped with two tops, the first method enables the processed crankshaft to have higher rigidity and adjusted eccentricity.

Due to the large manufacturing error of the chuck and the clamping error of the jaws, the eccentric chuck is not suitable for the crankshaft with higher precision requirement. When the diameter of the crankshaft is relatively thick and the eccentricity is not large, the crankshaft can be processed by the eccentric chuck, as shown in Figure 2.25. First, drill an eccentric center hole A for the main shaft journal on each of the two end faces, drill the eccentric center holes B, C, D, and E for each crank journal on one end face. Second, clamp and process the outer diameter, use the eccentric center hole A as the positioning datum on the eccentric chuck, the eccentric center hole on the other end face is supported with the top, and after aligning the datum journals at both ends with a dial indicator, turning the crank journal.

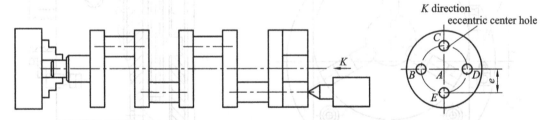

Figure 2.25　Clamping the workpiece with an eccentric chuck

Note: after turning the crank journal in one direction, use the center frame to support the crankshaft journal near the tailstock, loosen the chuck and the top, rotate the workpiece to support the the eccentric center hole with the top, make aligning, turning the other crankshaft journal.

Ⅲ. Crankshaft processing method

1. The sequence principle of rough and fine turning each journal(axial diameter)

For turning the crankshaft with small eccentricity, directly drill eccentric holes on the end face, clamping and turning can be carried out without other fixtures. For turning the crankshaft with large eccentricity and eccentric center hole on the end face which is difficult to be drilled, an eccentric fixture can be used.

The rigidity of the workpiece after clamping is different. In order to ensure the rigidity, the journal of the middle part of the crankshaft should be processed first, and then the crank journals on both sides should be turned. If the workpiece is long, the center frame support should be installed at the pre-processed journal. The diagram of clamping the center part of the main journal is shown as in Figure 2.26.

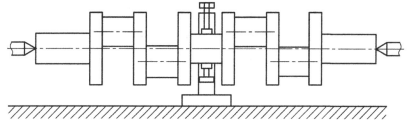

Figure 2.26   The diagram of clamping the center part of the the the main journal

2. Methods of reducing deformation of crankshaft during turning

When turning the crankshaft, the deformation of the workpiece always exists. The greater the deformation, the worse the quality of the workpiece. Therefore, reducing the deformation is the key to improving the quality of the crankshaft. During processing, as long as the deformation is correctly analyzed and solved, the deformation can be minimized and even be eliminated.

1) The main cause of crankshaft deformation

(1) The influence of difference of the workpiece in static balance on crankshaft deformation. During processing, the difference in static balance of the workpiece will produce centrifugal force, which will bend the rotation axis of the workpiece, making the turning depth of the outer circle unequal, thereby causing roundness error in the outer circle of the workpiece. The greater the difference in static balance, the greater the roundness error of the workpiece.

(2) The influence of the degree of looseness and tightness of the tops and supporting bolt on the deformation of the crankshaft. When machining crankshafts, especially slender crankshafts, if the tops or supporting bolt are too tight, the rotation axis of the workpiece will bend, and the parallelism error between the crankshaft journal axis and the supporting bolt axis will increase.

(3) The influence of incorrectly drilled center hole on crankshaft deformation. When machining the crankshaft, the incorrectly drilled center hole(that is to say, the center holes at both ends are not on the same line or the axis of the center holes at both ends is skewed), will cause the crankshaft deformation, making the part shake

slightly when rotating, which would result in high journal cylindricity error, and an increasing parallelism error between the crankshaft journal and the main shaft journal, and sometimes, deformation would damage the center hole and the tops, and even cause an accident.

(4) The impact of lathe accuracy and cutting speed on the deformation of the crankshaft. The worse the accuracy of the lathe, the greater the impact of the centrifugal force caused by the difference in static balance on the machining quality. The higher the cutting speed, the greater the centrifugal force, and the more serious the deformation of the workpiece.

2) Several methods to overcome crankshaft deformation

When turning crankshaft workpieces, the complex shape of the crankshaft and the long crank arm of it require that the overhang elongation of the turning tool should be extended. If not, the deterioration in the stiffness of the tool holder happens. These factors coupled with the impact load during cutting, result in low rigidity of the processing system and proneness to vibration and deformation of the tools. Therefore, it's necessary to improve the rigidity of crankshaft and cutting tools.

(1) Apply the fish maw shaped turning tool. The diagram of the fish maw shaped turning tool, whose protruding part is in fish maw shaped, is shown as in Figure 2.27, such a design not only can ensure that the crank does not collide with the tool holder when rotating, but also can improve the rigidity of the protruding part of the cutter body. The height h of fish maw is generally $(0.6-0.8)\,L$, the width of the cutter is B, and the height H of the clamping part of the cutter body should be as large as possible.

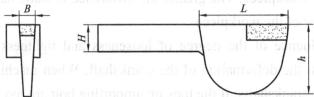

Figure 2.27   The diagram of fish maw shaped turning tool

(2) Apply the row type turning tool.The diagram of the row type turning tool is shown as in Figure 2.28. The structure of such a turning tool consists of a high-rigidity tool array and a tool head, and the supporting bolt used to fasten the tool head. This structure makes it more convenient to sharpen and replace tools, and saves the tool body material. However, when it is used for turning steel parts, the chip

removal is not smooth, in other words, chips are easy to squeeze into the gap between the front of the tool head and the tool row, which makes it not easy to observe the processing conditions. Therefore, this turning tool is mostly used for rough machining.

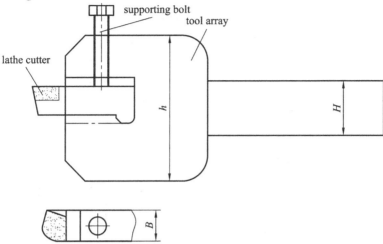

Figure 2.28   The diagram of the row-type turning tool

(3) Apply the auxiliary supporting turning tool. The diagram of the auxiliary supporting tool is shown as in Figure 2.29. Before processing parts, pre-processing a screw hole at the bottom of the lathe tool, and the supporting bolt is installed on the middle sliding plate, whose supporting length is adjusted and locked with a nut, such placement aims to increase the rigidity of the turning tool during rough machining. In this method, it is not convenient to rotate the tool holder.

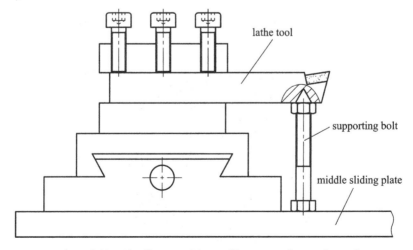

Figure 2.29   The diagram of the auxiliary supporting turning tool

(4) Improve the machining rigidity of the crankshaft. The diagrams of steps to

improve the machining rigidity of the crankshaft are displayed as Figure 2.30. As shown in Figure 2.30(a), install supporting bolts between the crank journal or the main journal; when the distance between the two crank arms is large, the distance can be supported by a harder wooden block or a stick, as shown in Figure 2.30(b); when the inner surface of the two crank is in an arc plane or inclined plane, the crank arms can be clamped on a pair of clamping plates, as shown in Figure 2.30(c);

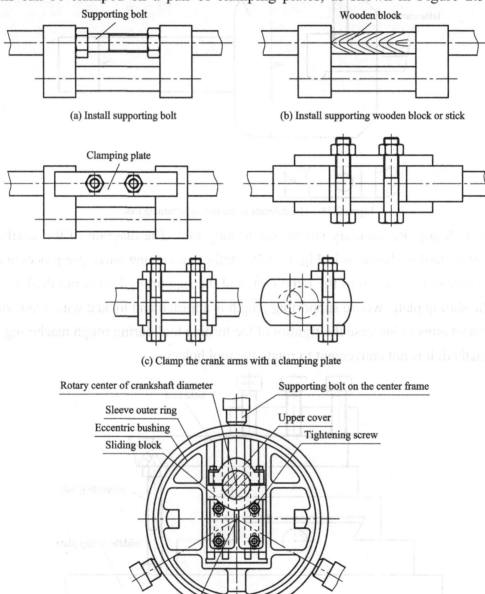

(a) Install supporting bolt

(b) Install supporting wooden block or stick

(c) Clamp the crank arms with a clamping plate

(d) Support the center frame and the eccentric bushing

Figure 2.30   The diagrams of steps to improve the machining rigidity of the crankshaft

when the long journal of the crankshaft is relatively large, in order to prevent the workpiece from vibrating, the center frame and eccentric bushing can be used for clamping, as shown in Figure 2.30(d). Before turning, lightly tighten the screws on the sliding block, then rotate the workpiece (or make the cutting tool work at low speed), adjust the supporting bolt on the center frame to align the outer circle of the eccentric sleeve, and then tighten the tightening screws.

3) Notes

(1) Take the process of drilling each center hole seriously, and make the center holes of the two ends be straight and on the same axis as far as possible.

(2) Carefully align the static balance of the workpiece, first use the two tops to gently support the crankshaft during machining, make the crankshaft start and stop at each rotation position. When the difference in static balance of the workpiece is found to increase after rough turning, it needs to be realigned, but to ensure that the workpiece has enough margin for finishing. When the margin of the raw material of the workpiece is too small, the static balance of the workpiece cannot be realigned, because it will cause waste due to the deformation of the workpiece.

(3) Properly tighten the center hole of the crankshaft during clamping, but not too tightly. If the clamping conditions permit, after clamping the end of the workpiece, the rear thimble support can be used to replace the outer circular support to avoid the negative influence of the clamping pressure.

(4) During turning, except for the parts of the workpiece that need to be turned, the rest should be supported by supporting bolts or pressed firmly with a pressure plate as much as possible, but it should not be too tight, otherwise, the workpiece will be deformed.

(5) When turning, the cutting speed should not be too high, and its range should not be too large.

(6) Pay attention to adjusting the gap of the lathe spindle, especially when the accuracy of the lathe is low.

IV. The measurement of crankshaft

1. The method of testing parallelism between crank journal and main shaft journal axis

The parallelism test diagram is shown as in Figure 2.31. First place the main journal of the crankshaft on the V-shaped frame, align it with a dial indicator so that the two ends of the main journal are at the same height, and then move the dial indicator to the crank journal to test whether the highest point values of each crank journal are the same.

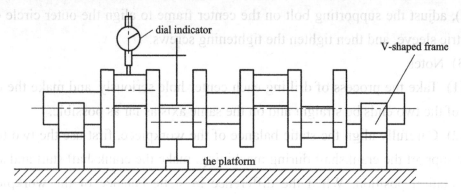

Figure 2.31   The diagram of testing the parallelism

2. The methods of testing the crank journal angle

1) Test the three-throw crankshaft when the angle of the crank journal is 120°

The diagram of testing the three-throw crankshaft with gauge block supported is shown as in Figure 2.32. First place both ends of the crankshaft on the V-shaped frame, adjust the distance between the supporting axes at both ends and the horizontal plane to be equal, and then put a gauge block under the crankshaft to make the line connecting the center of the crank journal and the center of the supporting journal to form an angle of 30° with the horizontal plane of the plate.

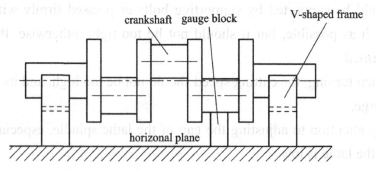

Figure 2.32   The diagram of testing the three-throw crankshaft with gauge block supported

The formula for the height of the gauge block is as follows:

$$h = A - \frac{1}{2}(D + R + d)$$

In this formula, h—the height of the gauge block, mm; A—the measured height between the supporting journal and the horizontal plane, mm; D—the diameter of the supporting journal, mm; R—the center distance between the supporting journal and the crank journal (eccentricity), mm; d—the diameter of the crank journal, mm.

When checking the included angle of the cranks, use a dial indicator, first measure the reading of the vertex of circular cone of one crank journal, and then measure the reading of the other crank journal. If the readings of the two are the same, then the included angle of the two crank journals is equal to $120°$. The crank angle measurement (when the included crank angle is $120°$) is shown as in Figure 2.33. If the readings of the two are different, the included angle of the cranks has an error. The following formula can be used to obtain the included angle θ and the included angle error $\Delta\theta$ between the connecting line of the center of the crank journal and the center of the supporting journal, and the horizontal plane of the plate with no gauge block supported.

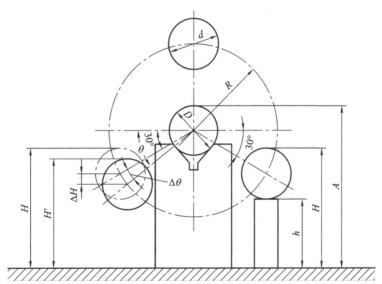

Figure 2.33   The crank angle measurement (when the included crank angle is $120°$)

$$\sin\theta = 0.5 - \frac{H' - H}{R}$$

In the formula, H'—the distance from the apex of the crank journal to the horizontal plane with no gauge block heightened, mm; H—the distance from the apex of the crank journal to the horizontal plane with gauge block heightened, mm;

Check the Trigonometric Function Table to find the included angle θ between the connecting line of the center of the crank journal and the center of the supporting journal, and the horizontal plane of the plate with no gauge block supported, then the crank angle error is $\Delta\theta = 30° - \theta$.

Example 2.2 Known that the supporting journal diameter of the three-throw crankshaft workpiece is $D = 60$ mm, the crank journal diameter $d = 55$ mm, the eccentricity $R = 50$ mm, the measured height of the supporting journal apex $A = 150.5$ mm, on the V-shaped frame, $H' = 122.5$ mm, $H = 130$ mm, please find the crank angle error.

Solution:

$$h = A - \frac{1}{2}(D+R+d) = 150.5 - \frac{1}{2}(60+55+50) = 68 \text{ mm}$$

$$\Delta H = H' - H = -0.5 \text{ mm}$$

then

$$\sin\theta = 0.5 - \frac{H'-H}{R} = 0.5 - \frac{-0.5}{50} = 0.5 + 0.01 = 0.51$$

as

$$\theta = 30°39'50''$$

So the angle error is

$$\Delta\theta = 30° - \theta = 30° - 30°39'50'' = -39'50''$$

It can be seen that the included angle $39'50''$ is smaller than $120°$, which is $119°20'10''$.

2) Measurement and check of other multi-throw crankshafts

In actual production, in addition to the three-throw crankshafts with all crankshaft journals equal mentioned above, various multi-throw crankshafts with unequal crankshaft journals will also need to be measured. The number of crankshafts may be four throws, six throws, etc. At this time, the following steps are needed to measure and check the included angle error.

(1) Measure the diameter of the supporting journal D and the diameter of each crank journal d, d_1, d_2, d_3, etc. and the center distance between the supporting journal and the crank journal R.

(2) Place the crankshaft on the V-shaped frame and adjust the two ends so that

its supporting axis is at the same height as the horizontal plane. Use a height vernier caliper to measure the distance between the horizontal plane and the outer circular point of the supporting journal (the measured height of the supporting journal apex)A.

(3) Calculate the distance between the apex of the crank journal and the horizontal plane H, H_1, H_2, H_3, etc. according to the following formula, and then use the height vernier caliper to correct the crank position according to the calculated value (the crank journal is heightened with a gasket block).

The measurement of crank angle and angle error when the included crank angle is $90°$ is shown as in Figure 2.34, when the included angle of the crank is $90°$,

$h = A - \dfrac{1}{2}(D+d) - 0.707R$, the center of the crank forms an angle of $45°$ with the horizontal plane.

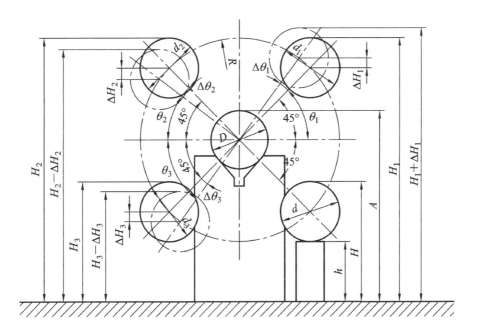

Figure 2.34   The crank angle and angle error measurement

(when the included crank angle is $90°$)

The measurement of crank angle and angle error when the included crank angle is $60°$ is shown as in Figure 2.35, when the included angle is $60°$, $h = A - \dfrac{D}{2} + \dfrac{d}{2}$, and the center of the crank is on the horizontal plane.

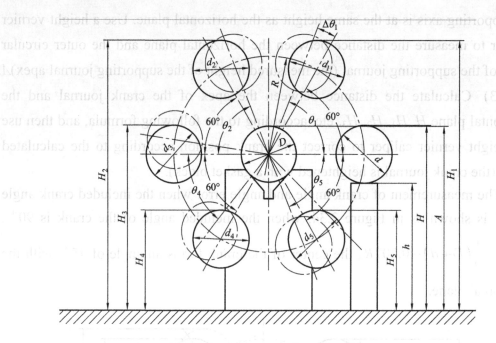

Figure 2.35　The crank angle and angle error measurement

(when the included crank angle is 60°)

The included angle and the included angle error is calculated as follows:

First, use a height vernier caliper to measure the distances between the apexes of each crankshaft (the diameter of each of them is d_1, d_2, d_3, d_4, d_5) and the horizontal plane as H_1, H_2, H_3, H_4, H_5.

Second, use the following formula to calculate the included angles between the line connecting the center of each crank journal and the center of the supporting journal, and the horizontal plane as θ_1, θ_2, θ_3, θ_4, θ_5.

① When the apex of the crank journal is higher than the apex of the supporting journal (i.e., $H_i > A$), there is

$$\sin \theta_i = \frac{H_i - A - \dfrac{d_i}{2} + \dfrac{D}{2}}{R}$$

② When the apex of the crank journal is lower than the vertex of the supporting journal (i.e., $H_i < A$), there is

$$\sin \theta_i = \frac{A - H_i - \dfrac{D}{2} + \dfrac{d_i}{2}}{R}$$

Third, simply convert the calculated included angles θ_1, θ_2, θ_3, θ_4, θ_5 to obtain the corrected included angles between the cranks, and get the angle error values $\Delta\theta_1, \Delta\theta_2, \Delta\theta_3, \Delta\theta_4, \Delta\theta_5$.

3. Eccentricity measurement

The crankshaft installed on the two tops is shown as in Figure 2.36. Measure H, h, d, d_1 with a dial indicator or a height gauge, and then use the following formula to calculate eccentricity.

$$e = H - \frac{d_1}{2} - h + \frac{d}{2}$$

where e—eccentricity, mm; H—the height of the highest point of the crank journal, mm; h—the height of the highest point of the main journal, mm; d—the diameter of the main journal, mm; d_1—the diameter of crank journal, mm.

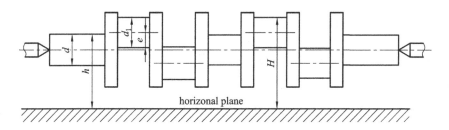

Figure 2.36   the crankshaft installed on the two tops

The skills training for crankshaft machining involves extensive knowledge, and it can be divided into two parts: crankshaft clamping and crankshaft machining.

【Skills Training】

Training 1

Aligning Four-throw Crankshaft and Drilling its Center Hole

I. Operation preparation

The preparation for aligning and drilling the center hole of the four-throw crankshaft according to its technique parameters in Figure 2.20 is shown in Table 2.7.

Table 2.7　The preparation for aligning and drilling the center hole of the four-throw crankshaft

No.	Name		Preparation
1	Material		45steelϕ52 mm $\times$ 350 mm
2	Device		CA6140 turning lathe, a four-jaw single-action chuck
3	Processing tools	Cutting tools	90° turning cutter, 45° elbow turning cutter, center drill A3.15/8 mm
4		Measuring tools	Vernier height scale 0.02 mm/(0-300 mm), vernier caliper 0.02 mm/(0-150 mm, 0-500 mm), magnetic seat dial indicator 0.01 mm/(0-30 mm)
5		Others	Processing soft claws, scribing rule, sample punch, scribing discs, square box, copper skin, flat-head screwdriver, adjustable wrench, top, drilling fixture, etc.

II. Operation procedure

The operation steps of aligning and drilling the center hole of the four-throw crankshaft is shown in Table 2.8.

Table 2.8　Operation steps of aligning and drilling the center hole of the four-throw crankshaft

No.	Operation procedure	Operation diagram
Step 1	Clamp the workpiece with 20 mm stretching out and align it	
	1) Turning the end face and drilling the center hole;　2) Turn the workpiece around and clamp it tightly, turning and drilling the center hole on the end face	
Step 2	Clamp the workpiece with the two tops	
	1) Rough turning the outer circle to ϕ 50 mm;　2) Semi-finish turning the outer circle to ϕ 48 mm	

Continued Table

No.	Operation procedure	Operation diagram
Step 3	Scribe reticle on the square box based on the ϕ 48 mm outer circle 1) Coat the surface of the workpiece with processing ink, scribe parts on a V-block, draw a cross center line on the end face, and prolong the line to the lateral generatrix of the outer circle; 2) Turn around the square box 90° and draw another cross line and lead the line to the lateral generatrix of the outer circle; 3) Draw the center hole lines of the crank journals	
Step 4	Clamp the workpiece with 20 mm stretching out with a four-jaw single-action chuck 1) Align the center lines of crank journals 1, 3, 2, and 4, and drill their holes respectively; 2) Turn around the workpiece, clamp it, and align the center hole lines of the crank journals, and drill the center holes on the end face	

Ⅲ. Workpiece quality test

(1) Get well prepared for aligning and scribing;

(2) Clamp the workpiece steadily, lines on the end face and the lateral generatrix of the outer circle must be clear and accurate.

Ⅳ. Notes

When the eccentricity of the workpiece is not large, it can be processed by the clamping method of one clamp and one top or two tops.

Processing Four-throw Crankshaft

I. Operation preparation

The preparation of processing the four-throw crankshaft is shown in Table 2.9.

Table 2.9   The preparation of processing the four-throw crankshaft

No.	Name		Preparation
1	Material		45 steel ϕ52 mm×340 mm
2	Device		CA6140 turning lathe, a four-jaw single-action chuck
3	Processing tools	Cutting tools	90° turning cutter, 90° outer circular turning cutter, 45° elbow turning cutter, right narrow turning cutter, $R2$ mm external arc turning tool, high-speed steel fine turning tool, external grooving tool, 60° external thread turning tool
4		Measuring tools	Vernier height scale 0.02 mm / (0–300 mm), vernier depth scale 0.02 mm / (0–200 mm), vernier caliper 0.02 mm / (0–150 mm, 0–500 mm), magnetic seat dial indicator 0.01 mm / (0–30 mm), M24 × 1.5—6G thread ring gauge
5		Others	Processing soft claws, scribing rules, sample punch, scribing discs, square box, copper skin, flat-head screwdriver, adjustable wrench, top, drilling fixture, and other common tools

II. Operation procedure

The operation steps of processing the four-throw crankshaft are shown in Table 2.10.

Table 2.10   Operation steps of processing the four-throw crankshaft

No.	Operation procedure	Operation diagram
Step 1	Support the center holes of crank journals 1, 2, 3, 4 with two tops.	
	Rough-turning the crank journals in the order 1, 2, 3 and 4 to size ϕ22 mm with length of ϕ18 mm	

Continued Table

No.	Operation procedure	Operation diagram
Step 2	Support the center holes of the main journals with two tops 1) Rough-turning the outer circle to $\phi29$ mm and with a length of 23 mm; 2) Rough-turning the outer circle to $\phi32$ mm and with a length of 18mm; 3) Rough-turning the main journals in the order 1, 2, 3 to size $\phi31$ mm and with a length of 18 mm	
Step 3	Turn around the workpiece and support the center holes of the main journals with two tops 1) Semi-finish and finish turning the main journal to $\phi30^{+0.015}_{+0.002}$ mm and with a length of 18mm; 2) Semi-finish turning and finish turning the crank journal to $\phi20^{0}_{-0.021}$ mm and with a length of $20^{+0.1}_{0}$ mm in the order of the middle first and then the two ends; 3) Break the sharp edges	
Step 4	Support the center holes of crank journals in the order 2, 3, 1, 4 with two tops Follow the order of 2, 3, 1, 4 for semi-finish turning and finish turning the crank journal to $\phi20^{0}_{-0.021}$ mm and with a length of $20^{+0.1}_{0}$ mm	
Step 5	Turn around the workpiece and support the center holes of the main journals respectively with two tops 1) Rough and fine turning the outer circle to $\phi27^{0}_{-0.021}$ mm and with a length of 24 mm; 2) Turn the undercut 3mm $\times$ $\phi21$ mm; 3) Turn the thread M24 $\times$ 1.5—6G; 4) Chamfer the sharp edges	

III. Workpiece quality test

The four-throw crankshaft should be tested according to its technical parameters in Figure 2.20.

1. Outer diameter and roughness of each part of the crankshaft

The main journal (on both sides) $\phi 30^{+0.015}_{+0.002}$ mm should be tested according to the dimensional tolerance, and it is unqualified if out of the tolerance.

The middle main journal (3 places) $\phi 20^{0}_{-0.021}$ mm should be tested according to the dimensional tolerance, and it is unqualified out of tolerance.

The crank journal (4 places) $\phi 20^{0}_{-0.021}$ mm should be tested according to the dimensional tolerance, and it is unqualified if out of tolerance.

The left keyway shaft diameter is $\phi 27^{0}_{-0.021}$ mm.

Roughness is $Ra1.6$ μm, or unqualified if downgrading.

2. Crank journal angle

The crank journal angle $90° \pm 15'$ for 4 places in total shall be tested according to the angle tolerance, and it is unqualified if out of tolerance.

3. Eccentricity and retract groove

The crankshaft is installed on the two tops to test the eccentricity (8 ± 0.0375) mm, 4 places in total. The eccentricity shall be tested according to the dimensional tolerance, and it is unqualified out of tolerance.

The retract groove is 3 mm × $\phi 21$ mm, its unmarked tolerance shall be tested according to GB/T 1804—2000.

4. Thread

The M24 × 1.5—6G thread is tested with the plug thread gauge. It is considered to be qualified if it passes the gauge and the stop, or unqualified if it fails to pass the gauge and the stop.

5. Length dimension

There are 7 places with a length of $20^{+0.1}_{0}$ mm and 7 places with a length of 35 ± 0.065 mm, which shall be tested according to the dimensional tolerance, and it is unqualified if out of tolerance.

There are 2 places with lengths of 24 mm, 44mm, 334mm, and 18mm, their unmarked tolerances shall be tested according to GB/T 1804.

6. Others

There are chamfers $C1$ mm, their unmarked tolerances shall be tested according to GB/T 1804—2000.

Roughness is $Ra3.2$ μm, or unqualified if downgrading.

IV. Notes

(1) When rough turning each journal, the round corner $R2$ mm should be processed with a margin.

(2) Observe during the turning process, and promptly check and correct any vibrations.

Module Three   Machining Gearbox

🅩 Task 1   Machining Gearbox Body

Subtask 1   Machining Gearbox Hole

【Knowledge and Skills Objectives】
Master the clamping, positioning and machining methods of gearbox.

【Related Knowledge】

I. Diagram of gear reducer

The diagram of gear reducer (also known as gear reducer box or gearbox) is shown as in Figure 3.1.

Here we introduce the machining of the gearbox, use gear reducer as an example, mainly discuss its gearbox hole. The D plane is the reference plane. The hole $\phi 40^{+0.025}_{0}$ mm has strict tolerance requirements benchmarked by the D plane, and the hole $\phi 40^{+0.025}_{0}$ mm is the reference hole for processing the surface of the gearbox. The distance between the axis of the hole $\phi 30^{+0.025}_{0}$ mm and the axis of the reference hole is (45 ± 0.05) mm, with the perpendicularity error between the two axes as 0.05 mm. The distance between the axis of the other hole $\phi 30^{+0.025}_{0}$ mm and the bottom plane is (115 ± 0.05) mm, the distance between the axis of the hole and the axis of the reference hole is (40 ± 0.05)

mm, with the perpendicularity error between it and the reference axis as 0.05 mm. The hole surface roughness of the three positions is $Ra1.6$ μm.

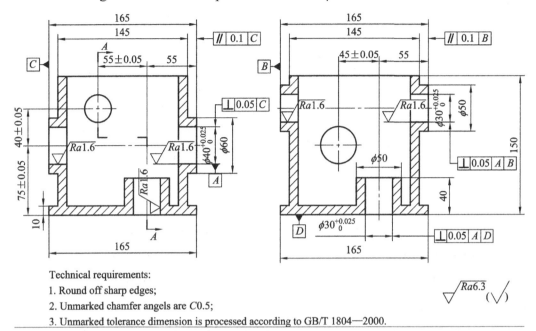

Technical requirements:
1. Round off sharp edges;
2. Unmarked chamfer angels are $C0.5$;
3. Unmarked tolerance dimension is processed according to GB/T 1804—2000.

Figure 3.1   The diagram of gear reducer

II. Clamping, positioning and machining the gearbox

1. Turn the vertical hole $\phi30^{+0.025}_{0}$ mm and the bottom plane 165 mm × 165 mm

After the workpiece is clamped on the faceplate angle iron, regard the side plane of the gearbox as the reference plane, turning the bottom plane in size 165 mm × 165 mm and the vertical inner hole $\phi30^{+0.025}_{0}$ mm.

2. Positioning the bottom plane, turning the through hole $2\times\phi40^{+0.025}_{0}$ mm

When turning the through hole $2\times\phi40^{+0.025}_{0}$ mm, first install the mandrel on the lathe spindle, and adjust the distance between the spindle axis and the angle iron plane to be 75 ± 0.05 mm. Second, install the gearbox, make the spindle pass through the horizontal blank hole $\phi40^{+0.025}_{0}$ mm of the gearbox, and align the outer side of the gearbox to be vertical to the flower disc with a square. Third, install the measuring rod on the bottom vertical hole $\phi30^{+0.025}_{0}$ mm, adjust the axis distance between the measuring rod and the spindle to be 45 ± 0.05 mm, then the through hole can be processed as in Figure 3.2, press the gearbox, remove the mandrel and the measuring rod.

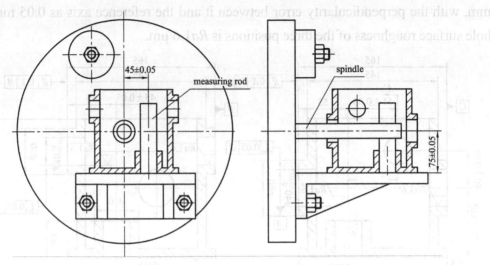

Figure 3.2   Diagram of turning the through hole $\phi 40^{+0.025}_{0}$ mm

3. Positioning the bottom plane, turning the through hole of $2 \times \phi 30^{+0.025}_{0}$ mm

When turning the through hole of $2 \times \phi 30^{+0.025}_{0}$ mm, first install the mandrel on the lathe spindle, and adjust the distance between the spindle axis and the angle iron plane to be 115 ± 0.05 mm. Second, install the gearbox, make the spindle pass through the horizontal blank hole $\phi 30^{+0.025}_{0}$ mm of the gearbox, and align the measuring rod of this hole to be vertical to the flower disc with a square. Third, install the measuring rod on the bottom vertical hole $\phi 30^{+0.025}_{0}$ mm, adjust the axis distance between the measuring rod and the spindle to be 55 ± 0.05 mm, as shown in Figure 3.3, press the gearbox, and remove the mandrel and the measuring rod.

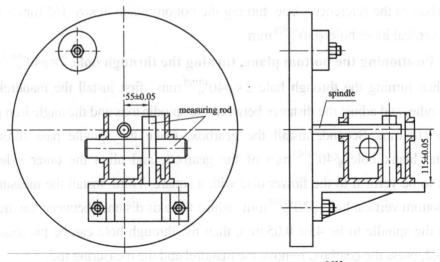

Figure 3.3   Diagram of turning the through hole $\phi 30^{+0.025}_{0}$ mm

【Skills Training】

Processing Gearbox Hole

Processing the gearbox hole of the gear reducer with its technique parameters in Figure 3.1.

I. Operation preparation

The preparation of processing the gearbox hole is shown in Table 3.1.

Table 3.1   The preparation of the processing the gearbox hole

No.	Name		Preparation
1	Material		45 steel plate, round steel(as blanks before welding) 165 mm × 165 mm × 15 mm(1 item), 145 mm × 145mm × 10 mm (2 items), 145 mm × 125 mm × 10 mm(2 items), ϕ65 mm × 15 mm(2 bosses), ϕ55 mm × 15 (2 bosses), ϕ50 mm × 30 mm (1 item), Note: weld above materials into a gearbox
2	Device		CA6150 turning lathe, faceplate, angle iron
3	Processing devices	Cutting tools	90° outer circular turning tool, 45° elbow turning tool, 75° inner hole turning tool, inner hole precision turning tool, two drilling bits ϕ27 mm and ϕ37 mm
4		Measuring tools	Vernier caliper 0.02 mm/(0–300 mm), micrometer 0.01 mm / (25–50 mm, 50–75 mm), vernier height ruler 0.02 mm / (0–300 mm), inner diameter dial indicator 0.01 mm / (18–35 mm, 35–50 mm), magnetic dial indicator 0.01 mm / (0–10 mm)
5		Others	Angle iron and pressure plate, screw, measuring rod, ruler, sample punch, scribing plate, square box, flat screw, adjustable wrench, drilling fixture, follow-rest, center rest, drilling chuck, adjustable wrench, fixed top, movable top, other common tools

II. Operation procedure

The operation steps of processing the gearbox hole are shown in Table 3.2.

Table 3.2   Operation steps of processing the gearbox hole

No.	Operation procedure	Operation diagram
Step 1	Scribe and drill the bottom hole $\phi27$ mm for processing the through hole $\phi30^{+0.025}_{0}$ mm and the bottom hole $\phi35$ mm for processing the through hole $\phi40^{+0.025}_{0}$ mm	
Step 2	Install the gearbox on a four-jaw single-action chuck, and turning the bottom plane and use the side face as the benchmark 1) Scribe and align the workpiece and clamp it; 2) Rough and fine turning the vertical bottom hole $\phi30^{+0.025}_{0}$ mm; 3) Turning both end faces to the height of 150 mm	
Step 3	Install the angle iron on the faceplate, insert the spindle into the mandrel hole, adjust the center height of the reference hole with a vernier caliper, gauge blocks, and a dial indicator, and turning the inner hole $\phi40^{+0.025}_{0}$ mm 1) Adjust the distance between the angle iron plane and the axis of the mandrel to 75 ± 0.05 mm; 2) Install the measuring rod in the vertical bottom hole $\phi40^{+0.025}_{0}$ mm to ensure that the center distance between the measuring rod and the mandrel is 45 ± 0.05 mm; 3) Take off the measuring rod and the mandrel; 4) Rough and fine turning the coaxial inner hole $\phi40^{+0.025}_{0}$ mm; 5) Turn the end faces of the hole $\phi40^{+0.025}_{0}$ mm to a length of 165 mm; 6) Chamfer $C1$ mm	45±0.05 75±0.05
Step 4	Turn the workpiece around, put the spindle into the mandrel; adjust the height of the reference hole center with vernier caliper, gauge and micrometer; turning the inner hole $\phi30^{+0.025}_{0}$ mm 1) Adjust the height between the angle iron and the spindle center to (115 ± 0.05) mm; 2) Insert the measuring rod in the vertical hole $\phi30^{+0.025}_{0}$ mm of the bottom face to ensure the distance between the measuring rod and mandrel to be 55 ± 0.05 mm; 3) Take off the measuring rod and the mandrel; 4) Turning the coaxial inner hole $\phi30^{+0.025}_{0}$ mm; 5) Turning the two end faces of hole $\phi30^{+0.025}_{0}$ mm and chamfer $C1$ mm	55±0.05 115±0.05

III. Workpiece quality test

The gear reducer should be tested according to its processing technical parameters as shown in Figure 3.1.

1. Inner holes and roughness

The inner holes $\phi 40^{+0.025}_{0}$ mm and $\phi 30^{+0.025}_{0}$ mm (horizontally and vertically) shall be tested according to the dimensional tolerance, and it is unqualified if out of tolerance.

Roughness is $Ra1.6$ μm, or unqualified if downgrading.

2. Center distance

The center distances 40 ± 0.05 mm, 75 ± 0.05 mm, 55 ± 0.05 mm and 45 ± 0.05 mm shall be tested according to the dimensional tolerance, and it is unqualified if out of tolerance.

3. Geometric tolerance

The parallelism (2 places) is 0.1mm, which shall be tested according to the geometric tolerance, and it is unqualified if out of tolerance.

The verticality (3 places) is 0.05mm, which shall be tested according to the geometric tolerance, and it is unqualified if out of tolerance.

4. Others

The overall size of the gearbox is 145 mm × 145 mm × 150 mm, their unmarked tolerance shall be tested according to GB/T1804—2000.

The end face dimensions of the hole are 165 mm(2 places) and 40mm, their unspecified tolerances shall be tested according to GB/T1804—2000.

The thickness of the bottom surface is 10 mm, its unmarked tolerance shall be tested according to GB/T1804—2000.

Other roughness is $Ra6.3$ μm , or unqualified if downgrading.

IV. Notes

(1) In order to ensure that the workpiece has sufficient machining allowance during processing, reasonable allowance allocation and scribing should be carried.

(2) If a certain hole is selected as the positioning datum, it is difficult to clamp the workpiece before machining other planes and holes. Therefore, the bottom face shall be selected as the positioning datum.

(3) The workpiece should be clamped firmly to prevent the throwing out to hurt people.

(4) The spindle speed should not be too high.

(5) The weld of the workpiece shall be retained.

Subtask 2　Measuring Gearbox Size

【Knowledge and Skills Objectives】

(1) Master the method of measuring the dimensions of various parts of the gearbox.

(2) Master the method of measuring the geometric tolerance of the gearbox.

【Related Knowledge】

I. Diagram of intersecting holes of the gearbox

The cross hole is identified according to the diagram of the gearbox case in Figure 3.1.

1. Vertical intersecting hole of the gearbox

The axes between the through holes $\phi 30^{+0.025}_{0}$ mm and $\phi 40^{+0.025}_{0}$ mm form the vertical intersecting hole structural line, which are perpendicular to each other but do not intersect.

2. Coaxial hole

The two through holes on the gearbox wall, $\phi 30^{+0.025}_{0}$ mm and $\phi 40^{+0.025}_{0}$ mm, are coaxial holes.

3. Three-dimensional intersecting hole

The hole $\phi 30^{+0.025}_{0}$ mm on the bottom plane, the holes $\phi 30^{+0.025}_{0}$ mm and $\phi 40^{+0.025}_{0}$ mm on the upper plane form a three-dimensional intersecting hole, which is perpendicular to each of the other two holes, but does not intersect with each of them.

II. Surface roughness measurement

The measurement of the surface roughness of the workpiece adopts the contrast method, and the standard gauge block is used to compare the processed surfaces to determine the *Ra* value.

Ⅲ. Measurement of dimensional accuracy

1. Hole inner diameter measurement

Vernier calipers are used to measure the hole inner diameter with low accuracy requirements. Inner diameter lever gauge, inner diameter micrometer, plug gauge, and caliper are used to measure the inner hole diameter with high accuracy requirements.

2. Hole depth measurement

Generally, a depth caliper is used to measure the hole depth, and a depth micrometer or gauge blocks are used to measure the hole depth with high accuracy requirements.

3. Center distance measurement

When measuring the distance between the vertical axis and the cross hole axis, measuring rod, dial indicator, gauge block, or micrometer are generally used.

4. Measuring the distance between two parallel holes

The diagram of measuring the distance between the two parallel holes is shown as in Figure 3.4. The distance *L* can be achieved by subtracting half of the addition of the size of gauge blocks from the outer diameter of the two measuring rods. The calculation formula is

$$L = L_1 - \frac{d_1 + d_2}{2}$$

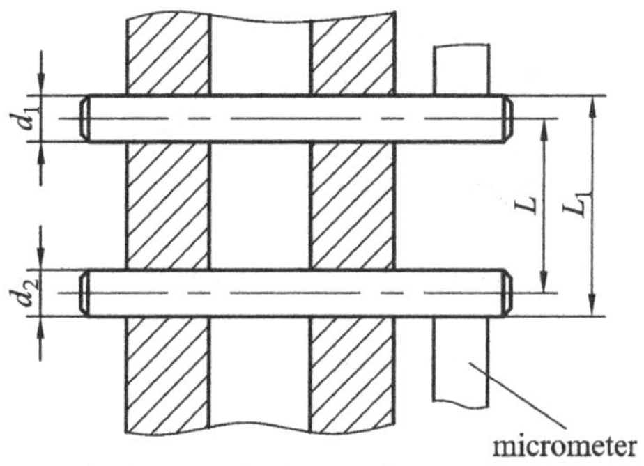

Figure 3.4   The diagram of measuring the distance between two parallel holes

5. Measuring the vertical intersecting hole distance

When the axes (shafts) of two holes are intersected vertically, the center distance of the two holes is measured by a measuring rod, a dial indicator, and gauge blocks. Place the gearbox on a plate, insert the measuring rod into the hole, and use a dial indicator to align the parallelism between the axis of the measuring rod and the plate. Place the gauge blocks on the surface of the plate, move the dial gauge to get the size of the measuring rod d_1, by making the gauge blocks and the dial gauge value consistent, as shown in Figure 3.5(a). Place another set of gauge blocks on the plate, move the dial indicator to get the size of the measuring rod d_2 by making the values of the gauge blocks and the dial indicator consistent, the size difference between the two sets of gauge blocks minus the difference of the two measuring rods is the actual distance between the axis of the two holes, as shown in Figure 3.5(b), The calculation formula is

$$L = L_1 - L_2 - \frac{d_1 - d_2}{2} \qquad (2.1)$$

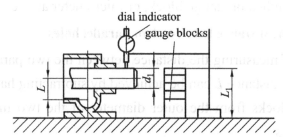

(a) The height of the measuring rod d_1 is consistent with that of the gauge blocks

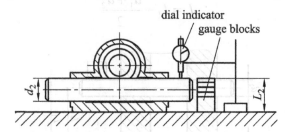

(b) The height of the measuring rod d_2 is consistent with that of the gauge blocks

Figure 3.5　The diagram of measuring the vertical intersecting hole distance

6. Measuring the distance between the axis of the hole and the datum plane

As shown in Figure 3.6, a dial indicator and gauge blocks are used to measure the hole axis and the datum plane. First, adjust the dial indicator value to zero scale; next, adjust the height of the gauge blocks and the distance between the plate (datum plane)and the bottom line of the inner hole, and then use the dial indicator to measure the height of gauge blocks. The numerical difference between the height of the gauge blocks and the height of the bottom line of the inner hole is the machining error between the axis of the hole and the datum plane.

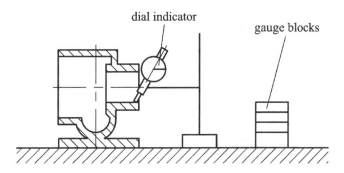

Figure 3.6   The diagram of measuring the distance between

the hole axis and the datum plane

Ⅳ. Geometric tolerance measurement

1. Coaxiality test

When the coaxiality of the hole is in a high requirement, a special measuring rod, as shown in Figure 3.7, can be used for accurately measuring the deviation value of straightness(coaxiality).

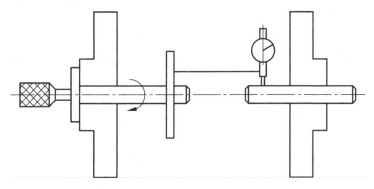

Figure 3.7   A special measuring rod used for coaxiality test

2. Measure parallelism of the axes of two holes

When the parallelism of the axes of two holes is not too high, a caliper can be used to measure the distance between the axes of two holes to make the parallelism test, as shown in Figure 3.8. When the parallelism of the axes of two holes is high, it can be measured with a micrometer, as shown in Figure 3.9.

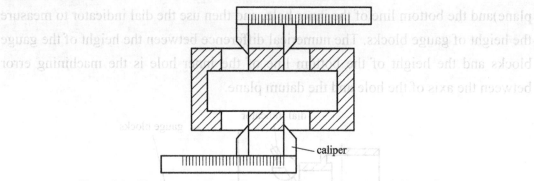

caliper

Figure 3.8   The diagram of parallelism test of the axes of two holes with a caliper

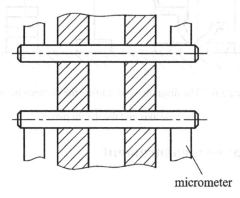

micrometer

Figure 3.9   The diagram of parallelism of the axes of two holes with a micrometer

3. Measure the parallelism of the hole axis to the datum plane

As shown in Figure 3.10, after inserting the measuring rod into the datum hole, a dial indicator is used to measure the parallelism of the axis of the hole to the datum plane. Place the workpiece on the plate, measure the two ends of the measuring rod with a dial indicator, and the value is the parallelism error of the axis of the hole to the datum plane.

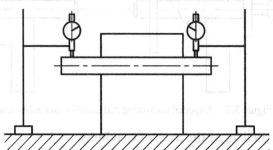

Figure 3.10   The diagram of measuring the parallelism of the hole axis

to the datum plane with a dial indicator

The measurement of parallelism of axes of two holes with a dial indicator is shown in Figure 3.11. First, align the datum axis to be parallel to the plate, and then measure the height of the two ends of the measuring rod. The measured difference is the parallelism of the axes of two holes.

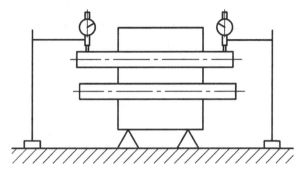

Figure 3.11   The diagram of measuring the parallelism of

the axes of two holes with a dial indicator

4. The perpendicularity of the end face to the hole axis

When measuring the perpendicularity of the end face to the axis of the hole, insert a measuring rod into the hole, install a dial indicator at one end of the measuring rod, make the measuring head of the dial indicator perpendicular to the end face of the hole, and rotate the measuring rod one round, and the perpendicularity (error) of the end face to the axis of the (inner) hole can be measured, as shown in Figure 3.12. If there is an inspection disc at one end of the measuring rod, insert the measuring rod into the hole to check the contact between the inspection disc and the end face by coloring method, or use a feeler gauge to check the gap between the inspection disc and the end face, as shown in Figure 3.13.

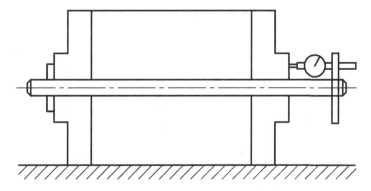

Figure 3.12   The diagram of measuring the perpendicularity of the end face to

the hole axis with dial indicator and measuring rod

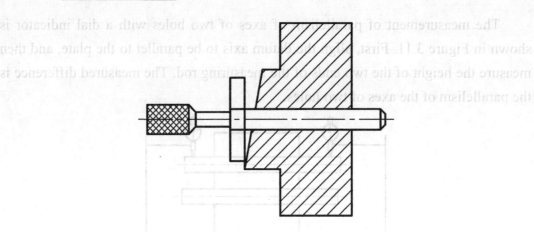

Figure 3.13   The diagram of measuring the perpendicularity of the end face to

the hole axis with measuring rod and inspection disc

V. Common errors in processing gearbox

The common errors in processing gearbox are shown in Table 3.1.

Table 3.3   Common errors in processing gearbox

Errors	Causes	Solutions
The inner hole is a horn hole	1) The axis of the spindle is not parallel to the movement of the saddle; 2) Bed saddle wear; 3) The turning tool wear happens midway	1) Check the spindle axis 2) Repair the saddle and rails 3) Choose right brand of turning tool, change the tool angle, and choose the cutting amount reasonably
The inner hole is not parallel to the bottom plane	Inaccurate bottom alignment	Repeatedly aligned
Surface roughness increased	1) Tool wear 2) Generate built-up tumor	1) Carefully sharpen the angle of the tool; 2) Change the turning speed
Center distance tolerance too large	Incorrect alignment of the workpiece	Continuously observe the scribing lines during processing and perform multiple measurements

【Skills Training】

Inspecting Gear Reducer Size

Inspect the size of the gear reducer according to its technique parameters shown in Figure 3.1.

Ⅰ. Operation preparation

The preparation for inspecting the size of the gear reducer is shown in Table 3.4.

Table 3.4　The preparation for inspecting the gear reducer size

No.	Name		Preparation
1	Material		Gear reducer
2	Device		Plate, square box
3	Processing tools	Measuring tools	A micrometer, a magnetic dial indicator 0.01 mm / (0−10 mm), a gauge block, a self-made measuring rod
4			
5		Others	A lift-jack

Ⅱ. Operation procedure

The operation steps for inspecting the size of the gear reducer (gearbox for short) are shown in Table 3.5.

Table 3.5　Operation steps of inspecting the gear reducer size

No.	Operation procedure	Operation diagram
Step 1	Place a plate for accuracy testing 1) Adjust the distance (75 ± 0.05) mm between the axis of the hole $\phi 40^{+0.025}_{0}$ mm and the datum plane, insert the measuring rod into the hole, and use a dial indicator to check the values at both ends of the measuring rod. 2) The actual size between the axis of the hole and the datum plane is the measured height subtracted by the radius of the measuring rod	 dial indicator　gauge block

No.	Operation procedure	Operation diagram
Step 2	Lift the bottom plane of the box with a lift-jack, and adjust the liftjack's height, make the distance be (40 ± 0.05) mm from the axis of the hole $\phi 40^{+0.025}_{0}$ mm to the axis of the hole $\phi 30^{+0.025}_{0}$ mm	dial indicator / gauge block
	1) Lift the box, align the parallelism of the axis of the hole $\phi 40^{+0.025}_{0}$ mm with a dial indicator, and place the gauge blocks, in order to make the size of the gauge blocks match the distance from the top of the measuring rod to the flat plate	
	2) Insert the measuring rod into the hole $\phi 30^{+0.025}_{0}$ mm , place another set of gauge blocks that have been calculated and combined on the plate (the same method as above), and the difference between the two sets of gauge blocks after each of them subtracted by the radius of each measuring rod is the axis distance	dial indicator / gauge block
Step 3	Adjust the distance of (45 ± 0.05) mm between axis of hole $\phi 40^{+0.025}_{0}$ mm and axis of the vertical hole $\phi 30^{+0.025}_{0}$ mm, and the distance of (55 ± 0.05) mm between the axis of hole $\phi 30^{+0.025}_{0}$ mm and axis of the vertical hole $\phi 30^{+0.025}_{0}$ mm on the bottom plane. Turn the workpiece over and support it with a lift-jack	dial indicator / gauge block
	1) Adjust the distance between the two axes to (45 ± 0.05) mm. Take the axis of the hole $\phi 40^{+0.025}_{0}$ mm as the datum plane, and insert the measuring rod into the holes $\phi 40^{+0.025}_{0}$ mm and $\phi 30^{+0.025}_{0}$ mm, test the parallelism of the axis of the hole $\phi 40^{+0.025}_{0}$ mm to the plate, place the gauge blocks to make the measuring rod consistent with the gauge blocks in height.	dial indicator / gauge block

Continued Table 2

No.	Operation procedure	Operation diagram
Step 3	Then use a dial indicator to measure the vertical hole $\phi 30^{+0.025}_{0}$ mm, place the gauge blocks to make the measuring rods consistent with the gauge blocks in height. The difference between two sets of gauge blocks after each of them subtracted by the radius of each measuring rod is the distance between the two axes; 2) Measure the distance of (55 ± 0.05) mm between two axes, its detection method is the same as above	
Step 4	Adjust the parallelism of the end faces of the two holes $\phi 30^{+0.025}_{0}$ mm and the parallelism of the end faces of the two holes of $\phi 40^{+0.025}_{0}$ mm The parallelism can be measured with a micrometer, and the error is accepted within 0.10 mm	micrometer
Step 5	Test the perpendicularity between the axis of the vertical hole $\phi 30^{+0.025}_{0}$ mm on bottom plane and the datum A, D. The two measuring rods are respectively inserted into the holes 1) Place the measuring rod in the hole $\phi 30^{+0.025}_{0}$ mm on the top support, align the parallelism between them with the dial indicator, rotate the box to make the dial indicator contact the measuring rod of the hole $\phi 30^{+0.025}_{0}$ mm on the bottom plane and record its value, rotate it $180°$ again under the condition that the dial indicator touches the measuring rod. The error of perpendicularity should not exceed 0.05 mm. 2) The method of measuring the verticality on the bottom plane and the axis of the vertical hole $\phi 30^{+0.025}_{0}$ mm is the same as above	dial indicator
Step 6	The method of measuring the perpendicularity error (0.05mm) between datum A and B, and the common axes of two horizontal holes $\phi 30^{+0.025}_{0}$ mm is the same as Step 5	

Ⅲ. Workpiece quality test

The gear reducer should be tested according to its technical requirements as shown in Figure 3.1.

1. Center distance

The center distance 75 ± 0.05 mm, 40 ± 0.05 mm, 45 ± 0.05 mm and 55 ± 0.05 mm should be accurately tested according to the dimensional tolerance, and it is unqualified if out of tolerance.

2. Geometric tolerance

The parallelism of 0.10 mm (2 places) and the perpendicularity of 0.05 mm(3 places) shall be tested according to the geometric tolerance, and it is unqualified if out of tolerance.

Ⅳ. Notes

(1) Do not damage the measuring rod during measurement.

(2) The plate should be wiped clean during testing to achieve the measurement accuracy.

Task 2　Machining Worm Gear Housing

【Knowledge and Skills Objectives】

(1) Master the scribing technology for worm gear housing;

(2) Master the selection of positioning datum in machining worm gear housing.

【Related Knowledge】

Ⅰ. Diagram of worm gear housing

A pair of worm gear and worm cooperate to form a reduction mechanism—worm gear housing(also known as worm gearbox), as shown in Figure 3.14. The worm gear is installed inside the worm gear housing, which has various shapes. Figure 3.15 shows the diagram of worm gear housing.

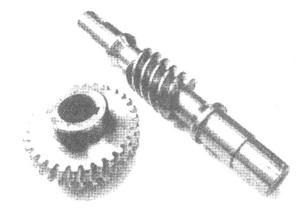

Figure 3.14   Picture of a pair of worm gear and worm (worm gear housing)

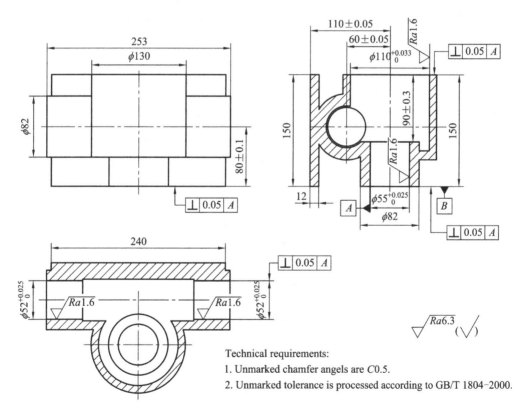

Technical requirements:
1. Unmarked chamfer angels are C0.5.
2. Unmarked tolerance is processed according to GB/T 1804-2000.

Figure 3.15   The diagram of worm gear housing

The dimensions of worm gear housing are introduced as follows:

(1) The axis of the worm gear housing hole $\phi 55^{+0.025}_{0}$ mm is used as the datum axis, and the distance between the datum axis and the bottom plane is (110 ± 0.05) mm.

(2) The perpendicularity error of the end faces of holes $\phi 55^{+0.025}_{0}$ mm and $\phi 110^{+0.033}_{0}$ mm to the datum axis is 0.05 mm.

(3) The end face of the hole $\phi 55^{+0.025}_{0}$ mm of the worm gear housing is used as the datum plane.

(4) The diameter of the inner hole of the worm is $\phi 52^{+0.025}_{0}$ mm, and the distance between the axis of the inner hole of the worm and the datum plane is (80 ± 0.1) mm.

(5) The distance between the axis of the worm and the axis of the worm gear is (60 ± 0.05) mm.

(6) The size of the worm gear housing is 253 mm × 150 mm.

(7) The size of the worm gear housing base is 150 mm × 240 mm.

II. Scribing the worm gear housing

(1) Scribe the worm gear housing. Place the worm gear housing on the plate, place a lift-jack on the base of the worm gear housing, and lift the worm gear housing.

Align the upper base line with the scribing disc. After aligning, draw a 12 mm line on two sides of the base, draw a distance of 110mm by a height ruler between the bottom base line and the axis of the worm, the distance between the axis of the worm gear hole and the axis of the worm is 60 mm, scribe the axes as shown in Figure 3.16.

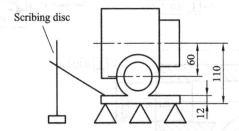

Figure 3.16   Scribing axial lines between the axis of the worm gear, and axis of

the worm and the upper base line of the worm gear housing

(2) Turn the workpiece over to scribe. Turn the workpiece 90° and use a lift-jack to support it.

Align the perpendicularity of the base and the axis of the worm gear hole with a 90° angle ruler, and at the same time, align the parallelism of the axes of the worm with scoring disc are perpendicular to the axis of the worm gear hole, draw a width line of the worm gear housing for 150 mm, and draw a 80 mm worm axis perpendicular to the axis of the worm gear hole, as shown in Figure 3.17.

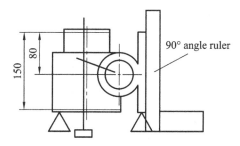

Figure 3.17   Scribing distance lines between the axis of the worm and the axis of the worm gear hole

(3) Turn the workpiece over to scribe. Turn the workpiece upright with a lift-jack support.

Align the base with a 90° angle ruler, align the the 12 mm between the upper line and the bottom line of the plate, and align the vertical line with a 90° angle ruler, draw a dimension line with a total length of 253 mm, pay attention to the symmetry of the dimension line and the worm gear hole axis, as shown in Figure 3.18.

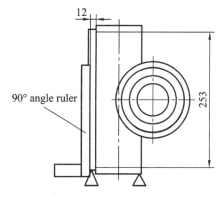

Figure 3.18   Scribing of the dimension line of the worm gear housing

Scribing is an important preparation work before processing the worm gear housing. By scribing, casting defects in the worm gear housing can be detected and corrected timely, which can ensure sufficient machining allowance and symmetrical dimensions.

III. Selection of positioning datum during processing the worm gear housing

Take the bottom plane of the base as the benchmark (positioning datum) when machining the holes of $\phi 55^{+0.025}_{0}$ mm and $\phi 110^{+0.033}_{0}$ mm in worm gear housing. Install the mandrel of the spindle, align the distance of (110 ± 0.05) mm between the angle iron and the axis of mandrel of the spindle. After the workpiece is installed, use the scribing plate to align the lines, as shown in Figure 3.19.

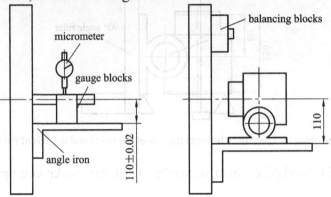

Figure 3.19   Machine holes benchmarked by the bottom plane

【Skills Training】

Processing the worm gear housing

Processing the worm gear housing with its technical parameters shown as in Figure 3.15.

I. Operation preparation

The preparation for processing the worm gear housing is shown in Table 3.6.

Table 3.6   The preparation for processing the worm gear housing

No.	Name		Preparation
1	Material		Box casting blank
2	Device		CA6150 lathe, faceplates, angle irons
3	Processing tools	Cutting tools	90° external turning tool, 45° turning tool, 75° inner hole turning tool, inner hole precision turning tool
4		Measuring tools	Vernier caliper 0.02 mm / (0–300 mm), micrometer 0.01 mm / (125 mm–150 mm), vernier height scale 0.02 mm/(0–300 mm), inner diameter dial indicator 0.01 mm/(50–160 mm), magnetic base dial indicator 0.01 mm/(0–10 mm)
5		Others	Clamping plate, screw, measuring rod, wrench, ruler, sample punch, scribing disc, square box, angle iron, slotted screwdriver, adjustable wrenches, tops and drilling fixture, and other common tools

Ⅱ. Operation procedure

The operation steps of processing the worm gear housing are shown in Table 3.7.

Table 3.7　Operation steps of processing the worm gear housing

No.	Operation procedure	Operation diagram
Step 1	Mill the bottom plane, use the bottom plane as the benchmark for turning and clamping	
Step 2	Install the workpiece by the faceplate and angle iron When machining the holes of $\phi 55^{+0.025}_{0}$ mm and $\phi 110^{+0.033}_{0}$ mm, take the bottom plane as the benchmark. Install the spindle axis, align the distance between the angle iron and the spindle axis (110 ± 0.05) mm, and align the workpiece after installation with the scribing disc	
Step 3	Place the gearbox on the angle iron and machine the worm gear hole 1) Install the gearbox based on the bottom plane, align the hole of $\phi 55^{+0.025}_{0}$ mm to make its end face parallel to the faceplate. 2) Processing the holes of $\phi 55^{+0.025}_{0}$ mm and $\phi 110^{+0.033}_{0}$ mm; 3) U-turn to machine the small end face of the hole $\phi 55^{+0.025}_{0}$ mm	
Step 4	Install the gearbox and machine the worm hole	

Continued table

No.	Operation procedure	Operation diagram
	1) Install the gearbox on the angle iron based on the bottom surface of the gearbox, align the center distance (50 ± 0.02) mm, insert the measuring rod into the hole $\phi 55^{+0.025}_{0}$ mm, and align the parallelism of the measuring rod and the faceplate to ensure the perpendicularity of the worm hole to the hole of $\phi 55^{+0.025}_{0}$ mm. Vertically align to test the center distance of (60 ± 0.05) mm between the worm gear hole and the worm hole, and align the distance of $(80 \pm 0.01$ mm between the datum plane and the axis of worm inner hole in parallel direction. 2) Machine the worm hole $\phi 55^{+0.025}_{0}$ mm and the other end face. 3) U-turn to processing the other end of the worm gear hole	

Ⅲ. Workpiece quality test

The worm gear housing should be tested according to its technical requirements as shown in Figure 3.15.

1. Installation of the faceplate and angle irons

Clamping should meet the safety requirements. Only safety is guaranteed, can we machine the workpiece. In particular, poor dynamic balance, looseness of counterweights, or bolts which can cause mechanical and quality accidents should be noticed.

2. Inner holes and roughness

The inner holes $\phi 55^{+0.025}_{0}$ mm , $\phi 110^{+0.033}_{0}$ mm and $\phi 52^{+0.025}_{0}$ mm should be tested according to the dimensional tolerance, and it is unqualified if out of tolerance.

Roughness is $Ra1.6\ \mu m$, or unqualified if downgrading.

3. Center distance

The center distance 60 ± 0.05 mm, 110 ± 0.05 mm, 80 ± 0.1 mm should be tested according to the dimensional tolerance, and it is unqualified if out of tolerance.

4. Geometric tolerance

The perpendicularity 0.05 mm (3 places) shall be tested according to the geometric tolerance, and it is unqualified if it is out of tolerance.

5. Others

The overall dimensions of the worm gear housing are 253 mm × 150 mm × 240 mm, ϕ 82 mm and ϕ 130 mm, their unspecified tolerance shall be tested according to GB1804—2000.

Distance between the datum plane and the axis of the worm inner hole (80 ± 0.1) mm shall be tested according to the dimensional tolerance, and it is unqualified if out of tolerance.

Roughness is $Ra1.6\ \mu m$, or unqualified if downgrading.

IV. Notes

(1) The 150 mm × 240 mm base is the datum plane for clamping, and the bottom surface plane should be turned or milled first.

(2) Before processing, it is necessary to do a good job in balancing the faceplate to avoid safety hazards such as vibration.

🅩 Task 3   Machining Bevel Gears

【Knowledge and Skills Objectives】

(1) Understand the impact of lathe accuracy on machining gearbox of bevel gears and common accuracy requirements for gearbox hole.

(2) Master the selection of positioning datum when machining gearbox hole.

【Related Knowledge】

I. Diagram of gearbox of bevel gears

Clamp and processing the diagram of gearbox of of bevel gears shown as in Figure 3.20.

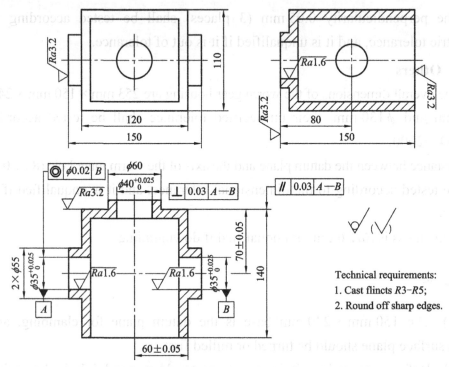

Technical requirements:
1. Cast flincts R3−R5;
2. Round off sharp edges.

Figure 3.20　The diagram of gearbox of bevel gears

The datum hole is an inner hole $\phi35^{+0.025}_{0}$ mm, and the distance between it and the top surface of one side is (70 ± 0.05) mm. The perpendicularity between the inner hole $\phi40^{+0.025}_{0}$ mm and the datum hole is required to be 0.03mm. In the top view, the parallelism between the upper and bottom end surfaces, and the datum $A—B$ is required to be 0.03mm. The two sides of the workpieces are distributed symmetrically, and the symmetrical size of one side is (60 ± 0.05) mm. Coaxiality error of the inner hole $\phi35^{+0.025}_{0}$ mm at both ends is within $\phi0.02$ mm.

II. Factors affecting the accuracy of gearbox holes

1. The radial circular runout error of the lathe spindle

The radial circular runout error of the lathe spindle causes roundness error of the bearing hole after turning and affects the surface roughness of the workpiece, as shown in Figure 3.21.

Figure 3.21   Lathe spindle radial circular runout error

2. The axial endplay of the lathe spindle

During turning, if the axial endplay of the lathe spindle changes, the bearing hole will not be perpendicular to the end face, as shown in Figure 3.22.

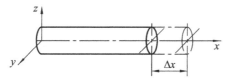

Figure 3.22   Axial endplay error

3. The influence of straightness error of the guide rail

The straightness error affects the form accuracy and position accuracy of the workpiece. The straightness error of the guide rail in the longitudinal vertical plane will change the position height of the tool in the longitudinal turning process, affecting the straightness of the workpiece and the cylindricity of the bearing hole.

III. Accuracy requirements of holes on the gearbox

The holes on the gearbox are usually bearing holes. The dimension accuracy, geometric accuracy and surface roughness of the bearing holes are required to be high. Otherwise, the matching accuracy of the bearing and the holes on the gearbox will be affected, the rotation accuracy of the axis will be reduced, and the transmission parts (such as gears) will easily produce vibration and noise.

Mutual position accuracy requirements of the main holes on the gearbox and the each side face are introduced as follows:

(1) Holes on the same axis shall have certain coaxiality requirements. The coaxiality tolerance of holes on the same axis is generally 0.01−0.04 mm.

(2) Mind the perpendicularity and parallelism between the axis of the bearing hole and the end face of the gearbox.

(3) Mind the perpendicularity when two box bearing holes cross each other.

IV. Principle of selecting positioning datum for gearbox with vertical intersecting holes

Due to the fact that there are multiple clamping processes during production, the principle of selecting positioning datum should be considered, it aims to make the mutual position accuracy requirements of holes and most machined surfaces with same positioning datum, so as to ensure that the assembly datum coincides with the design datum. Generally, a hole is machined based on its machined surface, and then other holes are machined based on this hole and their machined surface.

1. Selection of rough datum

(1) If it is necessary to ensure the position requirement between the machined surface and the non-machined surface on the part, the non-machined surface should be used as the rough datum.

(2) If one important surface of the part ensures uniform machining allowance, then it should be selected as the rough datum.

(3) When there are many machining surfaces on the part, in order to obtain sufficient machining allowance for each machining surface, the surface with the smallest machining allowance on the blank should be selected as the rough datum.

(4) The surface selected for the rough datum should be as flat and smooth as possible, free from burrs, risers, and other defects, in order to ensure accurate and reliable position accuracy.

(5) The rough datum should be avoided from repeated use, and it is allowed to be used once in the same dimension direction, otherwise, there will be significant positioning errors.

2. Selection of precision datum

(1) Principle of overlap datum: The design datum of the machined surface should be chosen as the precision datum as much as possible.

(2) Principle of unified datum: It is necessary to select as many machining surfaces as possible on the same set of precision machining parts to ensure the relative positional relationship between each machining surface.

(3) Principle of mutual datum: When the positional accuracy requirements between two machined surfaces on a part are relatively high, the method of repeated

machining with two machined surfaces as mutual datum can be adopted.

(4) Self-datum principle: Finish processing some surfaces which require small and uniform machining allowances, and often use the machined surface itself as the precision datum.

(5) Principle of easy clamping: The selected precision datum should ensure accurate and reliable positioning, simple clamping mechanism, and convenient operation.

【Skills Training】

Processing Gearbox of Bevel Gears

Processing the gearbox of bevel bears with vertical intersecting holes according to its technique parameters shown as in Figure 3.20.

Ⅰ. Operation preparation

The operation preparation for processing the gearbox of bevel gears is shown in Table 3.8.

Table 3.8　The preparation for processing the gearbox of bevel gears

No.	Name		Preparation
1	Material		Box casting blank
2	Device		CA6150 lathe, faceplates, angle irons, positioning sleeve
3	Processing tools	Cutting tools	90° external turning tool, 45° turning tool, 75° inner hole turning tool, drills of ϕ32 mm, ϕ37 mm
4		Measuring tools	Vernier caliper 0.02 mm / (0–150 mm), micrometer 0.01 mm / (2550 mm, 50 mm–75 mm), vernier height scale 0.02 mm / (0–300 mm), inner diameter dial indicator 0.01 mm / (18–35 mm, 35–50 mm), magnetic base dial indicator 0.01 mm / (0–10 mm)
5		Others	Angle iron and pressing plate, screw, spindle, wrench, scribing gauge, sample punch, measuring rod, marking disc, square box, slotted screwdriver, adjustable wrench, tops and drilling fixture, and other common tools

Ⅱ. Operation procedure

The operating steps of processing the gearbox of bevel gears are shown in Table 3.9.

Table 3.9   Operation steps of processing the gearbox of bevel gears

No.	Operation procedure	Operation diagram
Step 1	Scribe the gearbox 1) Scribe the top and bottom planes; 2) Scribe the central axis line of the reference hole $\phi 35^{+0.025}_{0}$ mm; 3) Scribe the central axis line of hole $\phi 40^{+0.025}_{0}$ mm	
Step 2	Install the faceplate, angle iron and spindle Align the angle iron so that it is parallel to the mandrel, and adjust the distance between the angle iron and the central axis of the mandrel to 55 mm	
Step 3	Clamp the gearbox(worm gear housing)	

Continued Table 1

No.	Operation procedure	Operation diagram
Step 3	1) Horizontally move the parts on the angle iron to roughly align the center of the worm gear housing; 2) Processing the bottom surface of the gearbox (bottom plane in the top view) to a size of 143mm (excluding the boss); 3) Fit the bottom plane of the part with the angle iron, and machine the distance of end sides of the gearbox to 110 mm to ensure even wall thickness of the gearbox (pay attention to aligning each machining face perpendicular to the axis of the lathe);	
Step 4	U-turn the workpiece, align the workpiece according to the scribing, extend and clamp the outer circle to ϕ60 mm	
	1) The total length of the lathe is 150 mm; 2) Turning the outer circle ϕ60 mm and the distance between two end faces to 140 mm; 3) Rough and fine turning inner hole $\phi 40_{0}^{+0.025}$ mm ; 4) Round off the sharp edges	
Step 5	Install the faceplate and angle iron, and align the height between the angle iron and the central axis	
	1) Install the core shaft(mandrel) in the spindle hole; 2) Place a gauge block on the angle iron. Align the distance between the alignment axis of mandrel and the angle iron by measuring the distance between the angle iron plane and the upper surface of the mandrel (the distance is equal to (70 ± 0.05) mm + the mandrel diameter/2 + the thickness of the positioning sleeve 15 mm)	

No.	Operation procedure	Operation diagram
Step 6	Clamp the workpiece with the positioning sleeve and turn the inner hole $\phi35_{0}^{+0.025}$ mm	
	1) Clamp the workpiece on the plane of the positioning sleeve benchmarked by the top surface; 2) Rough and fine turning the plane of the boss to ensure the alignment of 150 mm; 3) Through rough and fine turning inner hole $\phi35_{0}^{+0.025}$ mm (two holes), ensuring a distance of (70 ± 0.05) mm from the hole axis to the bottom plane; 4) Turn the side face to a length of 120 mm; 5) Turn around the workpiece and machine the other side to 150 mm; 6) Turning the other side face to a length of 120 mm	

III. Workpiece quality test

According to the technical requirements shown in Figure 3.20, inspect the gearbox of bevel gears.

1. Inner hole and roughness

The inner holes $\phi35_{0}^{+0.025}$ mm (2 places) and $\phi40_{0}^{+0.025}$ mm mm shall be all tested according to the dimensional tolerance, and it is unqualified if out of tolerance.

Roughness is $Ra1.6$ μm, or unqualified if downgrading.

2. The center distance

The center distance 70 ± 0.05 mm is tested according to the dimensional tolerance, and it is unqualified if out of tolerance.

3. Geometric tolerances

The perpendicularity is 0.03 mm, parallelism 0.03 mm, and coaxiality 0.02 mm, all of which are tested according to the geometric tolerance, or unqualified if out of tolerance.

4. Outline dimensions of the gearbox

The outline dimensions of the gearbox are 150 mm × 150 mm × 110 mm, their unmarked tolerance should be tested according to GB/T1804—2000.

IV. Notes

(1) The parallelism error between the end face of the positioning sleeve and the spindle axis should be controlled within 0.01 mm, and the perpendicularity error between the inner hole and the angle iron should be controlled within 0.01 mm.

(2) After installing the parts, make sure to find a good dynamic balance to avoid vibration of the machine tool.

(3) The parts must be securely installed to prevent loosening and safety accidents.

Module Four   Machining Assembly

⊿ Task 1   Machining Bisector Parts

Subtask 1   Machining Bearing Bush

【Knowledge and Skills Objectives】
(1) Understand the structure of bearing bush.
(2) Master the method of calibrating bearing bush.

【Related Knowledge】

Ⅰ. Diagram of bearing bush

Bearing bush is a kind of bisector part, the diagram of bearing bush is shown in Figure 4.1.

1. Size of bearing bush

(1) Bearing bush is a sleeve workpiece composed of two semicircles with the inner hole $\phi 60^{+0.027}_{0}$ mm and with 142 mm long. The oil groove is 105 mm long and with (6 ± 0.1) mm wide which is provided with an oil filling hole with a diameter $\phi (6 \pm 0.1)$ mm.

(2) The diameter of outer circles at both ends of the workpiece is $\phi 120$ mm and each with 10 mm long; the diameter of three steps on outer circle is $\phi 88^{0}_{-0.33}$ mm and each with 12 mm long; the diameter of two grooves on the outer circle is $\phi 80$ mm and with 43 mm long.

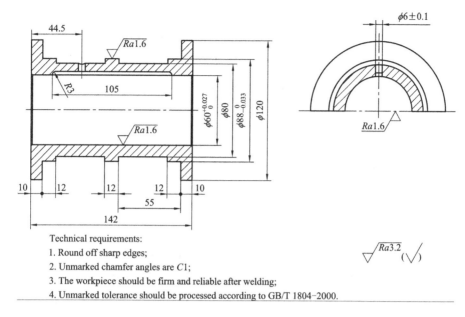

Technical requirements:
1. Round off sharp edges;
2. Unmarked chamfer angles are $C1$;
3. The workpiece should be firm and reliable after welding;
4. Unmarked tolerance should be processed according to GB/T 1804–2000.

Figure 4.1   The diagram of bearing bush

2. Concepts of bearing bush

The shaft matching the journal is called the bearing bush, the shaft supported by the bearing is called the journal, bearing bush is a part of sliding bearing.

Here we introduce the bearing bush, whose position in the sliding bearing is shown as in Figure 4.2. In order to improve the friction properties of the bearing surface, the antifriction material layer poured on the inner surface of the bearing is called bearing bush.

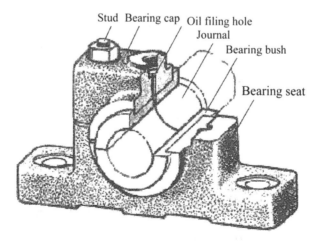

Figure 4.2   The position of the split bearing bush in the sliding bearing

Bearing bush, also known as bearing liner, is an important part in the sliding bearing. The materials of bearing bush and bearing liner are collectively called as bearing materials. Common sliding bearing materials include bearing alloy, wear-resistant cast iron, copper and aluminum based alloys, powder metallurgy materials, plastics, rubber, hardwood, graphite, polytetrafluoroethylene (Teflon, PTFE), Polyoxymethylene (POM), etc.

Bearing bush in the sliding bearing includes split structure and integral structure.

II. Methods for precisely correcting bisector

When the height gauges are used for alignment for the lateral generatrix on the two sides of the workpiece (each side uses one height gauge), and the workpiece is rotated to make the indicated value of the height gauge equal. Then rotate the workpiece by 180°, and use two height gauges to check whether the lateral generatrix on both sides of the workpiece are of the same height. If they are of different heights, it proves that the two half bearing shells are asymmetric, and it is necessary to continue to adjust the position of the workpiece.

The schematic diagram of precise correction of bisector is shown in Figure 4.3. When a scoring disc is used, for the lateral generatrix on both sides, turn around the scribing disc from the front to the end face, and then turn around the scribing disc to the back for aligning, as shown in Figure 4.3(a). After aligning, rotate the workpiece by 180° for aligning, as shown in Figure 4.3(b), until the scoring disc no longer needs to be adjusted.

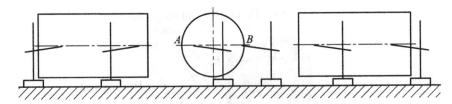

(a) The scribing disc is turn around from the front to the end face, and then turn around to the back for aligning

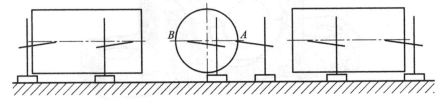

(b) The workpiece is rotated 180° for aligning

Figure 4.3   Schematic diagram of precise correction of bisector

【Skills Training】

Processing Bearing Bush

Processing the bearing bush according to its technique parameters requirements shown as in Figure 4.1.

Ⅰ. Operation preparation

The operation preparation for processing the bearing bush is shown in Table 4.1.

Table 4.1   The preparation for processing the bearing bush

No.	Name		Preparation
1	Material		Cast copper ϕ125 mm × 150 mm(Two pieces are welded into one)
2	Device		CA6150 lathe, four-jaw single-action chuck
3	Processing tools	Cutting tools	90° external turning tool, 45° elbow turning tool, inner hole rough and fine turning tool, grooving tool
4	Processing tools	Measuring tools	Feeler gauge, vernier caliper 0.02 mm / (0–150 mm), micrometer 0.01 mm / (50–75 mm, 75–100 mm), inner micrometer 0.01 mm / (50–160 mm)
5		Others	Scribing disc, slotted screwdriver, adjustable wrench, tops, and other common tools

Ⅱ. Operation procedure

The operation steps of processing the bearing bush is shown in Table 4.2.

Table 4.2   Operation steps of processing the bearing bush

No.	Operation procedure	Operation diagram
Step 1	Clamp the jaws of the four-jaw single-action chuck respectively on the vertical symmetrical bisector 1) Rough turning the outer circle and the end face; 2) After rough turning, the symmetrical bisector line shall be clear, and then align bisector line with the scribing disc again; 3) Rough and fine turning the inner hole $\phi60^{+0.027}_{0}$ mm	

Continued Table

No.	Operation procedure	Operation diagram
Step 2	Hold the workpiece with the top plate 1) Semi-finishing, finishing the outer diameter of ϕ120 mm and with the length of 10 mm on the right side; 2) Semi-finishing, finishing the outer diameter $\phi88_{-0.033}^{0}$ mm and with the length of 12 mm ; 3) Semi-finishing, finishing groove ϕ80 mm and with the length 43mm ; 4) U-turn the part for clamping, semi-finishing, finishing ϕ120 mm and with a length of 10 mm on the other side	
Step 3	U-turn and clamp the workpiece Turn end face and guarantee the total length to be 142 mm	

III. Workpiece quality test

The bearing bush shall be tested according to its technical parameters shown as in Figure 4.1.

1. Inner diameter and roughness

The inner hole $\phi60_{0}^{+0.027}$ mm shall be tested according to the dimensional tolerance, and it is unqualified if out of tolerance.

Roughness is Ra1.6 μm, or unqualified if downgrading.

2. The outer diameter and roughness

The outer diameter of $\phi88_{-0.033}^{0}$ mm shall be tested according to the dimensional tolerance, and it is unqualified if out of tolerance.

The outer diameters are $\phi 120$ mm and $\phi 80$ mm, their unmarked tolerance should be tested according to GB1804—2000.

Roughness is $Ra1.6$ μm, or unqualified if downgrading.

3. Length

The lengths are 10 mm (2 places), 12mm (3 places), 55 mm, 120 mm, 142 mm, their unmarked tolerance shall be tested according to GB1804—2000.

4. Others

Other part roughness is $Ra3.2$ μm, or unqualified if downgrading.

IV. Notes

(1) When clamping the workpiece on the four-jaw single-action chuck and aligning it with a scribing disc, it is necessary to align both the symmetrical bisectors on the end face of the workpiece, and make the symmetrical bisectors parallel to the bed guide rail.

(2) Processing the joint of bearing bushes into a flat surface by a milling machine, and then weld the workpiece together with welding materials.

(3) When aligning the workpiece, prevent the workpiece from falling off from the chuck.

Subtask 2   Machining Split Bearing Seat

【Knowledge and Skills Objectives】
(1) Master the method of processing split bearing seat.
(2) Master the method of measuring split bearing seat.

【Related Knowledge】

I. Diagram of split bearing seat

The diagram of split bearing seat (also known as split bearing housing)(a two half-body)is shown as in Figure 4.4.

(1) The bearing hole diameter is $\phi(80 \pm 0.02)$ mm with 45 mm long and the surface roughness of the bearing hole is $Ra1.6$ μm.

(2) The inner hole diameter is $\phi68$ mm with 53 mm long.

(3) The seal groove diameter is $\phi60$ mm with an angle of 40°, the groove width is 6 mm, and the groove bottom width is 2 mm.

(4) The shaft hole diameter is $\phi49$ mm with the surface roughness of $Ra1.6$ μm, and the two end faces distance 81 mm; the distance from the left end surface of the shaft hole to the center line is 40.5 mm.

(5) The distance from base datum to the center of the lower cover of the bearing seat is 65 ± 0.05 mm. On the upper cover and the lower cover of the bearing seat, there are respectively a boss and an inner step of 5 mm matched with each other.

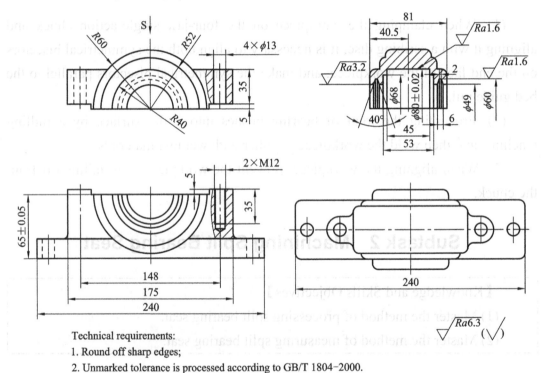

Technical requirements:
1. Round off sharp edges;
2. Unmarked tolerance is processed according to GB/T 1804-2000.

Figure 4.4　The diagram of split bearing seat

II. Processing technology for hole of the split bearing seat

Split bearing seat hole processing applies common turning. The following introduces the turning of mating split bearing seat hole by using disc and angle iron. The diagram of split bearing seat hole is shown as in Figure 4.5.

When machining the split bearing seat hole, the structure of the workpiece prevents the tool from turning directly from the outside. When turning, the middle bearing hole is invisible, so it is difficult to measure the middle bearing hole. According to the structural characteristics of the workpiece, it can be seen that the workpiece is composed of two parts, the upper cover and the lower seat. The upper cover can be removed during processing, so that the original invisible middle bearing hole is completely exposed, thus solving the difficulties in processing and measurement.

According to the structural characteristics of the workpiece and the analysis of the unfavorable and favorable conditions analyzed above, the following processing steps can be adopted.

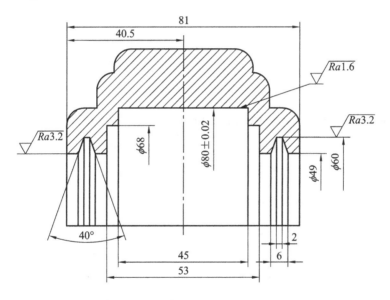

Figure 4.5   The diagram of split bearing seat hole

1. Clamping and aligning the workpiece

Place a layer of 0.1−0.2 mm thick gasket between the upper cover and the lower seat, select the bottom plane of the workpiece as the datum plane, and fasten it on the flower disc and angle iron with screws, as shown in Figure 4.6. Align the high and low positions of the workpiece according to the size requirements between the hole and the bottom plane, and correct the left and right positions of the workpiece according to the symmetry requirements of the positioning slots on the upper cover and the lower seat.

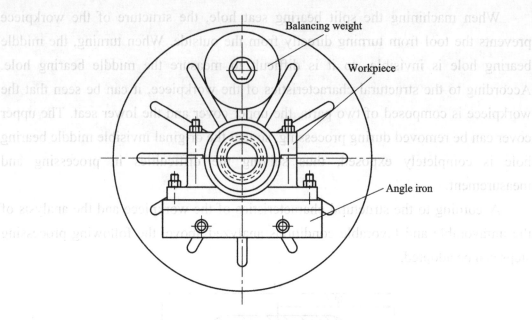

Figure 4.6   The diagram of clamping the workpiece on

the flower disc and angle iron with screws

2. Rough and fine turning the hole ϕ49 mm and the sealing groove ϕ60 mm

The bar of the lathe tool for turning the sealing groove ϕ60 mm should be as thick as possible, which is beneficial for the processing of the sealing groove.

3. Rough turning the bearing hole

Stop the lathe tool, remove the upper cover, and extend the turning tool to the position of the bearing hole (i.e $\phi(80 \pm 0.02)$ mm hole). Turn the flower disc by hand, and check whether the tool bar touches the hole ϕ 49 mm walls. If there is no interference, install the upper cover, and then rough turning a short distance at the bearing hole of the lower seat for measurement. According to the measured readings and the size requirements of rough turning, mark the degree on the scale of the middle sliding plate. At the same time, according to the requirements of the workpiece length, the reading of notches shall be recorded on the lathe saddle dial or marked on the tool bar. Then make the turning tool exit some in horizontal direction along the axis, install the upper cover, and rough turning the hole at one time according to the marks on the lathe saddle dial (or tool bar). Through the above steps, the coarse turning tool bar can be extended into the bearing hole ϕ80 mm in advance and thus the workpiece can be well processed without seeing the hole.

4. Finish turning the bearing hole

Similar to the rough turning method, a short distance is turned at the position of the bearing hole to measure the actual size of the hole diameter (using an inner caliper or inner micrometer. If the inner caliper or inner micrometer cannot directly extend into the inner hole during measurement, the upper cover can be opened and placed in the inner caliper or inner micrometer). When measuring, the upper cover can be installed to measure the difference between the actual size of the hole diameter and the required size, and then finish turning the inner hole. For some parts with high precision requirements, the method shown in Figure 4.7 can also be applied by using a dial gauge to achieve the back cutting amount. When in precision turning, considering the influence of "tool yield" of the tool bar, stop the turning to measure after a certain distance, or exit the turning tool and try to match it with the equipped bearing. If the equipped bearing can be gently pressed in by hand without looseness, it indicates that the hole diameter meets the requirements.

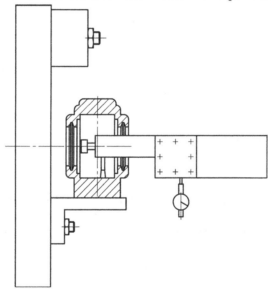

Figure 4.7   Method of using the dial indicator to achieve the back cutting amount

【Skills Training】

Turning Split Bearing Seat

Processing the split bearing seat according to its technique parameters shown as in Figure 4.4.

I. Operation preparation

The operation preparation for processing the split bearing seat is shown in Table 4.3.

Table 4.3   The preparation for turning the split bearing seat

No.	Name		Preparation
1	Material		Cast iron HT200
2	Device		CA6150 lathe, flower disc, angle iron
3		Cutting tools	Inner hole turning tool, 40° inner hole grooving cutter, flat bottom hole turning tool
4	Processing tools	Measuring tools	Vernier caliper 0.02 mm / (0–150 mm), micrometer 0.01 mm / (50–75 mm, 75–100 mm), inner micrometer 0.01mm / (50–160 mm), vernier height gauge 0.02 mm / (0–300 mm), magnetic base dial gauge 0.01 mm / (0–10 mm), inside caliper
5		Others	Screw, angle iron, scribing disk, slotted screwdriver, adjustable wrench, tops and drilling fixture, and other common tools

II. Operation procedure

The operation steps of processing the bearing seat are shown in Table 4.4.

Table 4.2   Operation steps of processing the split bearing seat

No.	Operation procedure	Diagram
Step 1	Clamp the workpiece on the angle iron. 1) Rough and fine turning the inner hole ϕ49 mm. 2) Turning the two end faces of the bearing hole to ensure the total length of 81 mm. 3) Turning the ϕ68 mm inner hole with 53 mm long	

Continued Table

No.	Operation procedure	Diagram
Step 2	Turning sealing groove 1) Use the grooving cutter to turning sealing groove on the wall of the hole ϕ49 mm into 6 mm deep. 2) Turning the sealing groove with a 40° inner hole grooving cutter at an angle of 40°	
Step 3	Turning the bearing hole Rough and fine turning the bearing hole $\phi(80 \pm 0.02)$ mm	

Ⅲ. Workpiece quality test

The split bearing seat is tested according to its technical parameters shown as in Figure 4.4.

1. Inner hole diameter and roughness

The inner hole $\phi(80 \pm 0.02)$ mm shall be tested according to the dimensional tolerance, and it is unqualified if out of tolerance.

The inner holes ϕ68 mm(2 places) and ϕ49 mm(2 places) shall be tested, their unmarked tolerance shall be tested according to GB/T1804—2000.

Roughness is Ra1.6 μm and Ra3.2 μm, or unqualified if downgrading.

2. Center distance dimension

The center distance is 65 ± 0.05 mm, which shall be tested according to the dimensional tolerance, and it is unqualified if out of tolerance.

3. Groove bottom dimensions

The groove bottom diameter ϕ49 mm shall be tested, its unmarked tolerance shall be tested according to the GB/T1804—2000.

The groove bottom width 2 mm and groove width 6 mm shall be tested, their unmarked tolerance shall be tested according to GB/T1804—2000.

The angle 40° of the sealing groove shall be tested, its unmarked tolerance shall be tested according to GB/T1804.

4. Length dimensions

Length dimensions such as 81mm, 40.5mm, 45mm and 53mm shall be tested, their unmarked tolerance shall be tested according to GB/T1804—2000.

5. Others

The other roughness is $Ra6.3$ μm, or unqualified if downgrading.

Ⅳ. Notes

(1) The distance from the center of the matching hole to the bottom plane is (65 ± 0.05) mm. If the contact plane of the upper cover and lower seat is still used as the datum plane, the dimensional accuracy of (65 ± 0.05) mm will not be guaranteed, affecting the assembly and use of the workpiece. Therefore, the position of the mating hole must be corrected by regarding the workpiece's bottom plane as the datum plane, and the bisection requirements of the contact plane between the upper cover and the lower seat must be guaranteed when processing the workpiece's bottom plane and the sealing groove on the upper milling machine, so that the processing of the mating hole can meet the assembly and use requirements.

(2) First processing the hole $\phi 49$ mm and the sealing groove $\phi 60$ mm, then turn the bearing hole $\phi(80 \pm 0.02)$ mm. In this way, the tool bar used for turning bearing hole shall be thicker and the rigidity of the tool bar would be increased.

📀 Task 2　Machining Mold

Subtask 1　Machining Cylinder Mold

【Knowledge and Skills Objectives】

(1) Understand the design requirements for the cavity of the mold.

(2) Master the method of turning cylinder mold.

【Related Knowledge 】

I . Cylinder and cylinder mold diagrams

The diagram of cylinder is shown in Figure 4.8, and the diagram of cylinder (casting) mold is shown in Figure 4.9, its parameter requirements are as follows:

(1) The cylinder mold is a clamping mold composed of two blocks, with an overall dimension 220 mm × 150 mm × 40 mm, two blocks are materials of the same size.

(2) In the cylinder mold, the length of the cylinder body is 138 mm, diameter $\phi 38 \pm 0.05$ mm, surface roughness $Ra0.8$ μm. The riser is 36 mm long and with an angle of $10°$.

(3) Diameters of two oil holes are $\phi (20 \pm 0.03)$ mm, surface roughness $Ra0.8$ μm. The two oil holes are connected with the oil pump body with the casting fillet $R2$ mm.

(4) There are 4 pin holes on the mold, with a diameter $\phi 16$ mm and center distance of each of mating holes 130 mm × 90 mm.

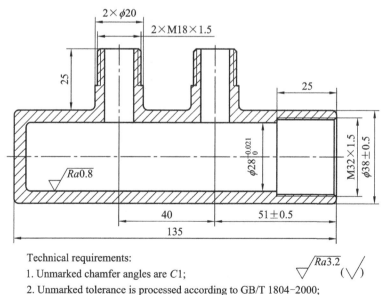

Technical requirements:
1. Unmarked chamfer angles are $C1$;
2. Unmarked tolerance is processed according to GB/T 1804−2000;
3. Casting fillet $R2$.

Figure 4.8   The diagram of cylinder

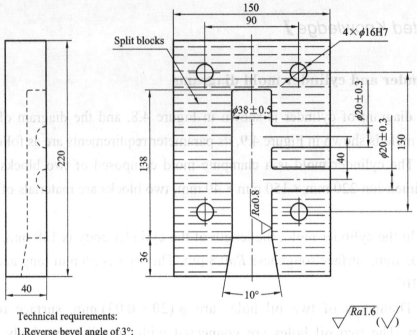

Technical requirements:
1.Reverse bevel angle of 3°;
2. Unmarked tolerance is processed according to GB/T 1804−2000;
3. Casting fillet R2.

(a) The mold plan view

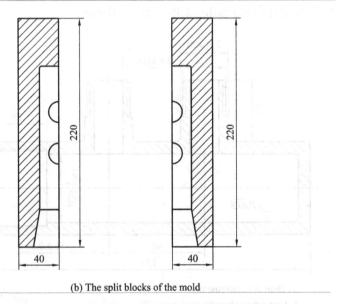

(b) The split blocks of the mold

Figure 4.9　　The diagram of cylinder mold

Ⅱ. Method of Processing Cylinder Mold

The cylinder is made of aluminum, whose temperature is very high after heating, so the material of cylinder mold is the hot work die steel H13. The cylinder mold adopts a split form, after the casting is completed and the casting is cooled, the cylinder mold is easy to be taken out. Use the positioning pin to combine the two blocks of mold and install the combination on the four-jaw single-action chuck (noted shortly as chuck) for processing. The specific perpendicularity error of the mold shall be less than 0.05 mm, the flatness error of the mold shall be less than 0.03 mm, and the surface roughness $Ra1.6$ μm. After the blocks are combined, the datum plane should be selected, and the center line and vertical line should be drawn on the plate for alignment as Figure 4.10.

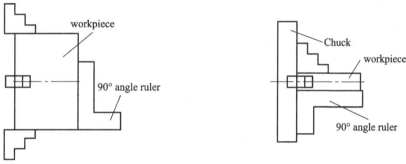

(a) Using a 90° angle ruler to find the perpendicularity between the end face and the spindle axis
(b) Using a 90° angle ruler to align the parallelism of the side surface and the spindle axis

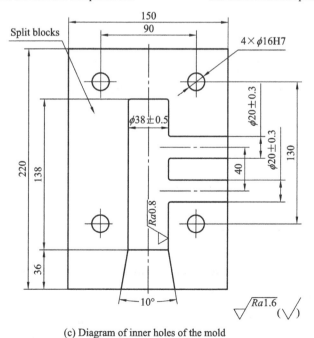

(c) Diagram of inner holes of the mold

Figure 4.10 The alignment of the workpiece

After the blocks are combined, the perpendicularity between the end face and the main spindle axis can be aligned with a dial indicator or a 90° angle ruler, as shown in Figure 4.10(a). The parallelism between the side surface and the main spindle axis can be aligned by 90° angle ruler shown in Figure 4.10(b).

As shown in Figure 4.10(c), after machining and aligning the workpiece, drill the hole of ϕ35 mm, rough and finish turn the inner hole of the mold to size $\phi(38 \pm 0.5)$ mm, fillet $R2$ mm, and with the length of 174 mm, turning riser angle 10° and the riser with 36 mm long. U-turn the workpiece by 90°, find out the error between the center line of the oil hole and the main spindle axis, control it within 0.3 mm, and turning two oil holes $\phi(20 \pm 0.03)$ mm respectively.

After the oil hole is turned, the length of the oil hole cannot be controlled to preferred value, and one processing plug needs to be installed in the hole, its diagram is shown as in Figure 4.11. The inner arc (as the casting fillet) $R3$ mm is processed on the end face of the processing plug, and the workpiece cools quickly after casting, so the casting fillet of the processing plug becomes larger. The outer diameter surface is knurled with straight knurling to facilitate exhaust, and it should have interference fit with the inner hole $\phi(20 \pm 0.03)$ mm.

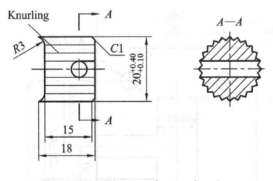

Figure 4.11   Diagram of processing plug

III. Design requirements for cavity

Cavity is the part of the mold where the workpiece (blank) is formed, and plays the role of holding the material. After the cavity is processed, the surface of the inner wall of the cavity should be polished and grounded to make the surface roughness $Ra0.8$ μm or lower. The draft angle is generally $3° \sim 5°$. If the angle is too large, the material is wasted, and if the angle is too small, it is difficult for material returning.

IV. The mold processing procedures

(1) The size of each surface of the workpiece is processed by a milling machine, and the perpendicularity error is controlled within 0.05 mm. The upper and lower dies are equipped with direct pins to connect the two blocks together for drilling and reaming. On the plate, mark the cross-center line and the axis of the oil outlet $\phi 20$ mm for alignment.

(2) After the workpiece is clamped, it needs to be aligned, and the datum plane is required to be perpendicular to the axis of the spindle and parallel to the chuck surface.

(3) Processing the cylinder body first, then process the oil hole, and check the perpendicularity of the axis of the main spindle and the axis of the end face before processing the oil hole.

(4) Machine the surface roughness of the cavity to $Ra0.8$ μm, quenching the mold before opening it, manually polish each surface so that the surface roughness reaches $Ra0.4$ μm with the draft angle of $3°-5°$.

【Skills Training】

Machining Cylinder Casting Mold

Processing the cylinder (casting) mold according to its technique parameters shown in Figure 4.9.

I. Operation preparation

Operation preparation for machining cylinder casting mold is shown in Table 4.5.

Table 4.5   Operational preparation for machining cylinder casting mold

No.	Name		Preparation
1	Material		H13, 2 items of 220 mm × 150 mm × 40 mm
2	Device		CA6150 lathe
3	Processing tools	Cutting tools	90° turning tool, 45° turning tool, inner hole coarse turning tool, inner hole fine turning tool, drill bit $\phi 18$ mm, drill bit $\phi 35$ mm, drill bit $\phi 5.8$ mm, straight knurled cutter $m0.5$, reamer $\phi 6H7$

Continued Table

No.	Name	Preparation
4	Measuring tools	Vernier caliper 0.02 mm / (0–300 mm), micrometer 0.01 mm / (0–25 mm), caliper, 90° angle ruler (0–300 mm), magnetic base dial indicator 0.01 mm / (0–10 mm)
5	Others	Scribing disc, slotted screwdriver, adjustable wrench, tops and drilling fixture, and other common tools

II. Operation procedure

Operational steps for machining cylinder casting mold are shown in Table 4.6.

Table 4.6   Operational steps for machining cylinder casting mold

No.	Operation step	Diagram
Step 1	After assembling the workpiece, install it on the four-jaw single-action chuck	
	1) Align the end face of the workpiece with a 90° angle ruler; 2) Align the side face of the workpiece with a 90° angle ruler	
	3) Drill cylinder hole to size ϕ35 mm and with 173mm long; 4) Rough and fine turning the cylinder inner hole to size $\phi(38 \pm 0.5)$ mm with fillet $R2$ mm; 5) The length of the riser is 36 mm and with an angle of 10°	
Step 2	Turn around the workpiece 90°, align and clamp it	
	1) Again, drill hole ϕ18 mm and connect it with the cylinder block; 2) Rough and finish turning the inner hole to diameter $\phi(20 \pm 0.3)$ mm; 3) The inner hole fillet is $R2$ mm	

Continued table

No.	Operation step	Diagram
Step 3	Move the workpiece up by 40 mm，align and clamp it 1) Drill hole ϕ18 mm and connect it with the cylinder block; 2) Rough and finish turning the inner hole to diameter $\phi(20 \pm 0.3)$ mm; 3) The inner hole fillet is $R2$ mm	
Step 4	Manufacture processing plug 1) Extend the workpiece to 30 mm, and clamp the end face; 2) Rough and finish turning the outer circle to size of $\phi20^{+0.1}_{-0.2}$ mm and a length of 20 mm; 3) Processing straight knurling $m0.5$ and a length of 20 mm, and cut off at the outer length of 19 mm; 4) Turn the workpiece round and clamp it, turning the end face at the length of 18 mm; 5) Chamfer $C1$ mm	
Step 5	Drive the processing plug into the hole ϕ20 mm, align it with the end face of the mold, and open the mold 1) Polish the cavity surface and fillets to make the surface roughness $Ra0.8$ μm or lower; 2) Drill holes ϕ5.8 mm at 25 mm deep (2 places) at 8mm from the end face of the processing plugs, ream the holes ϕ6 mm, insert the straight pin, and the end face of the straight pin should be lower than the outer surface of the processing plug	
Step 6	Test the mold Press the mold together, pour paraffin wax, open the mold after the paraffin wax cools, and check the dimensions of each part of the casting	

III. Workpiece quality test

The cylinder mold and processing plug shall be tested according to their technical parameters requirements shown in Figure 4.9 and Figure 4.11.

1. Cylinder mold

1) Inner diameters and roughness

The inner diameters $\phi(38 \pm 0.5)$ mm and $\phi(20 \pm 0.3)$ mm (in two parts) shall be tested according to the dimensional tolerance, and it is unqualified if out of tolerance.

The riser's angle is $10°$, its unmarked tolerance shall be tested according to GB/T 1804—2000.

Roughness is $Ra0.8$ µm, or unqualified if downgrading.

2) Lengths and distance of hole center lines

The lengths are 36 mm and 138 mm, their unmarked tolerance shall be tested according to GB/T1804—2000.

The distance of hole center lines is 40 mm, its unmarked tolerance shall be tested according to GB/T 1804—2000.

3) Others

Roughness in other parts is $Ra1.6$ µm, or unqualified if downgrading.

2. Processing plug

The external diameter $\phi 20^{+0.4}_{+0.1}$ mm shall be tested according to the dimensional tolerance, and it is unqualified if out of tolerance.

The length dimensions are 15 mm and 18mm, their unmarked tolerance shall be tested according to GB/T1804—2000.

The outer circular pattern of the processing plug shall be clear, otherwise, it is unqualified.

IV. Notes

(1) Use the milling machine to processing the dimensions of each surface of the workpiece, and the perpendicularity error shall be controlled within 0.05 mm.

(2) After being clamped, the workpiece needs to be aligned. The datum plane is required to be perpendicular to the spindle axis and parallel to the chuck surface.

(3) Before the oil hole is processed, the perpendicularity between the main spindle and the end face shall be checked.

(4) Processing the surface of the cavity to $Ra0.8$ μm. Quenching the mold before opening it, manually polish each surface to $Ra0.4$ μm with draft angle of $3°-5°$.

Subtask 2   Machining Gear Mold

【Knowledge and Skills Objectives】

(1) Understand the design requirements for the cavity of mold.

(2) Master the method of turning and forging mold.

【Related Knowledge】

I . Diagram of spur gear and diagram of the spur gear mold

The diagram of spur gear is shown in Figure 4.12. Spur gears are processed in batches. The diagram of spur gear mold is shown in Figure 4.13.

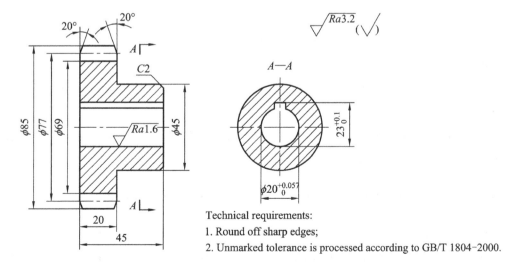

Figure 4.12   The diagram of spur gear

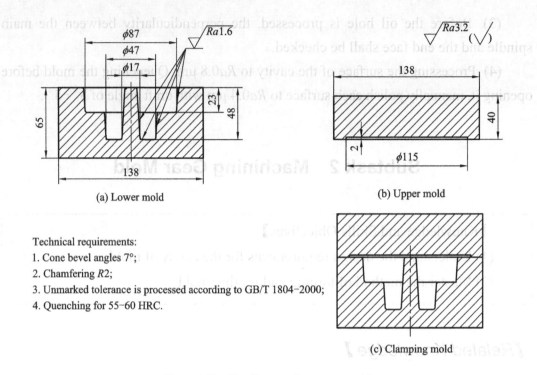

(a) Lower mold

(b) Upper mold

Technical requirements:
1. Cone bevel angles 7°;
2. Chamfering R2;
3. Unmarked tolerance is processed according to GB/T 1804−2000;
4. Quenching for 55−60 HRC.

(c) Clamping mold

Figure 4.13   The diagram of spur gear mold

(1) The spur gear mold is a closed one, composed of two blocks, the size of the lower mold is 138 mm × 138 mm × 65 mm, and the size of the upper mold is 138 mm × 138 mm × 40 mm.

(2) The major end diameter of the cone of the lower mold is ϕ17 mm, the bevel angle of the cone is 3.5°, and the length of the cone is 48 mm.

(3) The inverted diameter of the cone of the lower mold is ϕ47 mm, and the bevel angle of the cone is 3.5°.

(4) The inverted diameter of the cone of the lower mold is ϕ87 mm, the bevel angle of the cone is 3.5°, and the length of the cone is 23 mm.

(5) Turning the annular flash groove on the end face of the upper mold with a diameter ϕ115 mm and a depth 2 mm.

Ⅱ. Test requirements for processing the spur gear mold

After the gear is heated, whose temperature is very high, so the material of the mold is 5CrMnMo, which is a hot forging mold material. The mold adopts a split form, and the roughness of each machined surface in the upper mold is Ra1.6 μm to facilitate the return of material.

After each item is processed, test the combination of the parts of the mold, pour paraffin into the mold, wait for the paraffin to cool, and then open the mold to observe whether the paraffin can be released as a whole. If the mold cannot be released, check whether the draft angle of the cavity wall is too small or the surface roughness is too high.

Ⅲ. Design requirements for processing the cavity of mold

The cavity is the part of the mold where the workpiece is formed, and it plays the role of holding the material. After the mold cavity is processed, the surface of the inner wall of the cavity should be polished and grounded to make the surface roughness reach $Ra1.6$ μm or lower. The draft angle is generally 3° to 7°. If the angle is too large, the material is wasted, and if the angle is too small, it is difficult to return the material.

【 Skills Training 】

Machining Spur Gear Mold

Processing the spur gear mold according to its technical parameters shown in Figure 4.13.

Ⅰ. Operation preparation

Operation preparation for machining the spur gear mold is shown in Table 4.7.

Table 4.7 Operational preparation for machining spur gear mold

No.	Name		Preparation
1	Material		5CrMnMo raw material, 138 mm × 138 mm × 65 mm and 138 mm × 138 mm × 40 mm
2	Device		CA6150 lathe
3	Processing tools	Cutting tools	90° turning tool, 45° turning tool, internal threading tool
4		Measuring tools	Vernier caliper 0.02 mm/ (0−300 mm), micrometer 0.01 mm/ (0−25 mm), caliper, 90° angle ruler (0−300 mm), magnetic base dial indicator 0.01 mm/ (0−10 mm)
5		Others	Scribing disc, slotted screwdriver, adjustable wrench, tops and drilling fixture, etc.

Ⅱ. Operation procedure

Operational steps of machining spur gear mold are shown in Table 4.8.

Table 4.8   Operational steps of machining spur gear mold

No.	Operation procedure	Diagram
Step 1	Scribe the lower mold and then align it after installation 1) Align the workpiece with a 90° angle and turning the end face; 2) Turning the annular flash groove with the inner hole to the size of ϕ84 mm with a groove width of 19 mm, and a length of 22 mm; 3) Turning the inner hole of the inner groove to the size of ϕ45 mm with a bottom mold cone of ϕ19 mm and a length of 44mm (see left); 4) Turning the inverted cone to the size of ϕ87 mm and length 23 mm, and the size of ϕ47 mm, with a bottom mold cone of ϕ17 mm and length 48 mm, and then round corners R2 mm (see right); 5) Polish each surface to Ra1.6 μm	
Step 2	Install and align the workpiece 1) Turning the annular flash groove to the size of ϕ115 mm with a depth of 2 mm, and then round corner R2 mm; 2) Polish the surface to Ra1.6 μm	
Step 3	Test the mold Pour the paraffin into the lower mold, pour out the paraffin after the paraffin is cooled, and measure the dimensions of each part	

III. Workpiece quality test

The spur gear mold is tested according to its technical requirements shown in Figure 4.13.

1. Inner diameters and roughness

The internal diameters are $\phi 87$ mm, $\phi 47$ mm and $\phi 115$ mm. The unmarked tolerance shall be tested according to GB/T 1804—2000.

Surface roughness of the lower mold is $Ra 1.6$ μm, or unqualified if downgrading.

2. Others

The round corner is $R2$ and the draft angle is $7°$, the unmarked tolerance shall be tested according to GB/T 1804—2000.

IV. Notes

In the process of machining the spur gear mold, it is necessary to mainly master the machining methods for cavity and draft angle, and the polishing and sharpening methods for the mold body.

Task 3   Machining Shaft and Kit Combination

Subtask 1   Machining Tri-eccentric Shaft Sleeve

【Knowledge and Skills Objectives】

(1) Be able to analyze and solve the quality problems in machining eccentric combination workpiece.

(2) Master the methods of machining eccentric combination workpiece.

【Related Knowledge】

I . Diagram of combination workpiece of eccentric shaft sleeve

The assembly diagram of combination workpiece of eccentric shaft sleeve is shown as in Figure 4.14.

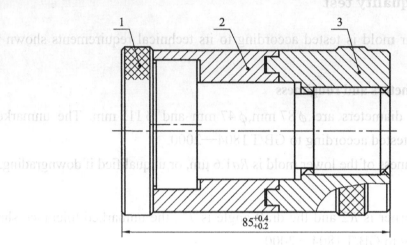

1-Eccentric shaft; 2-Eccentric sleeve; 3-Thread sleeve.

Figure 4.14   The assembly diagram of eccentric shaft sleeve

The combination workpiece of eccentric shaft sleeve consists of 3 items, including eccentric shaft, eccentric sleeve and thread sleeve, the diagram of each of them is shown in Figure 4.15, Figure 4.16 and Figure 4.17. After the three items are assembled together, the contour of the workpiece should be straight without protrusions. After assembly, the total length of the workpiece is required to be $85^{+0.4}_{+0.2}$ mm.

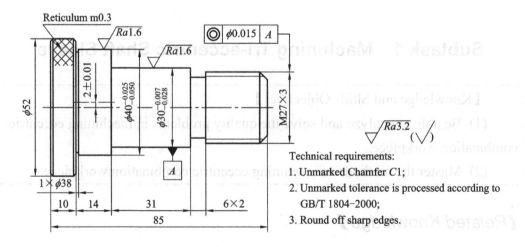

Technical requirements:
1. Unmarked Chamfer $C1$;
2. Unmarked tolerance is processed according to GB/T 1804-2000;
3. Round off sharp edges.

Figure 4.15   Diagram of item 1(eccentric shaft)

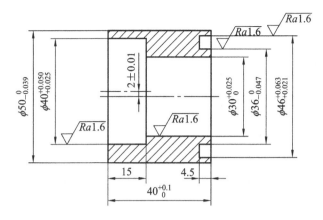

Technical requirements:
1. Unmarked Chamfer $C1$;
2. Unmarked tolerance is processed according to GB/T 1804–2000;
3. Round off sharp edges.

Figure 4.16   Diagram of item 2(eccentric sleeve)

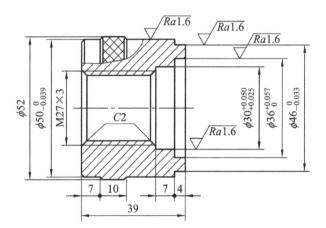

Technical requirements:
1. Unmarked Chamfer $C1$;
2. Unmarked tolerance is processed according to GB/T 1804–2000;
3. Round off sharp edges.

Figure 4.17   Diagram of item 3 (thread sleeve)

II. Analysis on processing technology of combination workpiece

1. Analyze and solve the quality problems in machining combination workpiece

1) Phenomenon of one high and one low on the outer surface of the workpiece

When two (or more than two) items are combined, the phenomenon of "one high and one low" appears on the outer surface of the workpiece, as shown in Figure 4.18. This is caused by improper clamping or processing. Important fitting surfaces should be machined in one clamping. For some ready-to combined workpieces, we can also leave a machining allowance on their outer surface, and the outer surface can be processed uniformly after the combination. At this time, the linearity of the lateral generatrix on the outer circular side of the workpiece is much better.

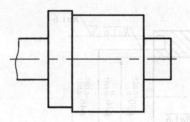

Figure 4.18   The phenomenon of one high and one low on the outer surface of the workpiece

2) Gap between two end faces

The reasons for the gap between the two end faces (see Figure 4.19) after the workpiece is combined are as follows:

(1) The datum axis of the workpiece and the lathe axis are not coaxial, as shown in Figure 4.19(a). Two-tops turning can be used to ensure the coaxiality of the two axes.

(2) Due to the error in the clamping, the parallelism error between two end faces is large, as shown in Figure 4.19(b).

Before processing the workpiece, it is necessary to carefully align the parallelism between the datum axis of the workpiece and the axis of the lathe.

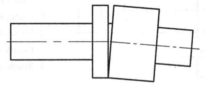

(a) The coaxiality error between the datum axis of the workpiece and the lathe axis

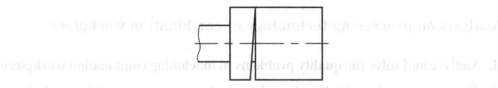

(b) The parallelism error between two end faces

Figure 4.19   Gap between the two end faces

2. Solve the key problems when turning combination workpiece

The key step in turning combination workpiece is to determine the datum part.

1) Determine the datum part

Carefully analyze the assembly relationship of the combination, determine the datum part, which directly affects the position accuracy of the parts after their assembly. If a part has an assembly relationship with multiple parts, then this part is usually the datum part, as shown in Figure 4.15—the eccentric shaft of Item 1.

2) Turning the datum part

When the combined workpiece is processed, the datum part should be turned first, and then other parts of the workpiece should be turned in turn according to the order of the assembly.

3) Precautions when turning datum part

(1) The dimensions (radial dimension and axial dimension)that affect the fitting accuracy of the parts should be processed to reach the middle value of the limit size as far as possible, and the processing error should be controlled within the middle value allowed in the diagrams of items. The geometric error and position error of each surface should be minimized as far as possible.

(2) For the fitting of cones, the cone angle error of the cone should be small. When turning, the height of the turning tool tip should be equal to the height of the axis of the workpiece to avoid hyperbola error.

(3) For the fitting of eccentric parts, the eccentricity of the eccentric parts should be consistent, and the axis of each of the eccentric parts should be parallel to the datum axis of the part.

(4) For thread fitting, the pitch diameter size of the external thread should be controlled at the minimum limit size, and that of the internal thread should be controlled at the maximum limit size to make the fit gap become slightly larger.

3. Guarantee the assembly accuracy

When turning the remaining parts of the workpiece, on the one hand, it should be carried out according to the technical requirements of the datum part, and on the other hand, it should be adjusted according to the actual measurement results of the processed datum parts and other parts, and make full use of fitting during turning, research for fitting, combined processing and other means to ensure the assembly accuracy for the combination workpiece.

4. Processing methods and principles of the combination workpiece

According to the technical requirements and structural characteristics of each part and the technical requirements of the workpiece, the processing method of each part, the processing times of each main surface (according to choice of rough turning, semi-finishing and finishing turning) and processing sequence are formulated. Usually, the datum surface should be machined first. The machining principles are as follows:

(1) Fitting of internal and external threads. Generally, the external thread is used as the datum part to be processed first, and then the internal thread is processed, because the size of the external thread is easy to measure.

(2) Fitting of the inner and outer cones. The outer cone is used as the datum part to be processed first, and then the inner cone, such a sequence is good for controlling the size.

(3) Fitting of eccentric parts. The datum part is regarded as the eccentric shaft, which is good for easy inspection. The eccentric sleeve and other parts are machined according to the assembly sequence. When the inner and outer eccentric parts are processed, the same clamping method is applied to ensure the eccentricity accuracy of the fitting parts.

III. Measurement knowledge

The eccentric values of eccentric shaft and eccentric sleeve are often measured with a pendulum meter or a magnetometer base with dial indicator, the pictures of two instruments are shown as in Figure 4.20.

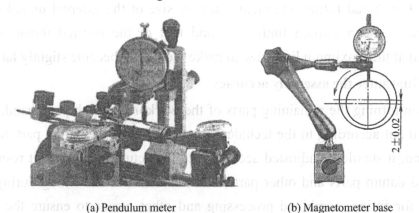

(a) Pendulum meter　　　　　　　(b) Magnetometer base

Figure 4.20　Instruments to measure eccentric values

When measuring, rotate the workpiece by hand to check the circular runout error, observe the value change of the dial indicator, so as to determine the eccentricity value. When the eccentricity value is 2 ± 0.02 mm, the displayed value of the dial indicator should be $2 \times (2 \pm 0.02)$ mm, that is, 4 ± 0.04 mm. If the swing range of the small dial indicator pointer exceeds 4 mm, the swing range of the large dial indicator pointer is qualified to be ± 0.04 mm. As shown in Figure 4.21, if the tolerance range is 0.08 mm, it is not acceptable.

±0.04

Figure 4.21   Tolerance of the swing range of the large dial indicator pointer

【Skills Training】

Processing the Tri-eccentric Bushing

Processing the combination workpiece of the tri-eccentric bushing according to technique parameters each of the combination workpiece shown in Figure 4.15 to Figure 4.17.

Ⅰ. Operation preparation

The operation preparations for machining each workpiece of the tri-eccentric bushing is shown in Table 4.9.

Table 4.9   Operation preparation for machining each
workpiece of tri-eccentric bushing

No.	Name		Preparation
1	Material		Medium carbon steel ϕ55 mm × 200 mm
2	Device		CA6150 lathe
3	Processing tools	Cutting tools	90° turning tool, 45° turning tool, inner hole coarse turning tool, inner hole fine turning tool, inner and outer 60° thread turning tool, cutting tool, m0.3 reticulate knurling tool, drilling bit ϕ28 mm, drilling bit ϕ34 mm, drilling bit ϕ22 mm, A2.5/6.3 center drill, end face grooving cutter
4		Measuring tools	Vernier caliper 0.02 mm / (0−150 mm), micrometer 0.01 mm / (25−50 mm), magnetic base dial indicator 0.01 mm / (0−10 mm), M27 thread ring gauge, M27 × 3 thread ring gauge, M27 × 3 thread plug gauge
5		Others	Scribing disc, slotted screwdriver, adjustable wrench, tops and drilling fixture, etc.

II. Operation procedure

The operation steps for machining the tri-eccentric bushing are shown in Table 4.10.

Table 4.10   Operation steps for machining the tri-eccentric bushing

No.	Operation step	Diagram
	Item 1(datum part)	
Step 1	Extend the workpiece by 95mm and clamp it	
	1) Turning the end face; 2) Rough and fine turning the outer circle to size $\phi52$ mm and with a length of 90 mm; 3) Rough turning the outer circle to size $\phi45$ mm and with a length of 75 mm; 4) Rough turning the outer circle to size $\phi31$ mm and with a length of 61 mm; 5) Rough and finish turning the outer circle to size $\phi27_{-0.4}^{-0.2}$ mm and with a length of 30 mm	
	1) Knurl $m0.3$ at the outer circle of $\phi52$ mm; 2) Chamfer $C1$ mm; 3) Turning undercut 6 mm × 2 mm, chamfer end face, turning thread M27; 4) Finish turning the outer circle $\phi30_{-0.028}^{-0.007}$ mm, and round off sharp edges	
Step 2	Loosen the workpiece, align the eccentricity and clamp the item 1 tightly	
	1) Align the out circle to size $\phi45$ mm in eccentricity 2 ± 0.01 mm, and align the lateral generatrix of the outer circle to size $\phi30_{-0.028}^{-0.007}$ mm with the dial indicator, make it parallel to the axis of the lathe; 2) Turning the outer circle of the eccentric shaft to size $\phi40_{-0.050}^{-0.025}$ mm; 3) Round off sharp edges; 4) Turning 1 × $\phi38$ mm undercut, for smooth edge; 5) Cut off the extra after chamfering, ensure the length of 86 mm	

Continued Table 1

No.	Operation step	Diagram
Step 3	U-turn and clamp the workpiece after aligning the $\phi 30$ mm outer circle tightly by padding copper sheets	
	1) Turn end face to guarantee a total length of 85 mm; 2) Chamfer $C1$ mm	
	Item 2	
Step 1	Extend the workpiece by 50 mm and clamp it	
	1) Turning the end face; 2) Drilling the hole of $\phi 28$ mm and with a length of 45 mm; 3) Rough and finish turning the outer circle to size $\phi 50_{-0.039}^{0}$ mm and with a length of 41 mm; 4) Chamfer end face $C1$ mm; 5) Cut off the part to guarantee the length of 42 mm	
Step 2	1) Finish turning the end face to guarantee a total length of 41 mm; 2) Rough and finish turning the inner hole to size $\phi 30_{0}^{+0.025}$ mm ; 3) Groove the end face; 4) Round off sharp edges of the outer circle	
Step 3	U-turn the workpiece and clamp it after aligning	
	1) Finish turning the end face to guarantee a length of $40_{0}^{+0.1}$ mm; 2) Clamp the workpiece with copper sheets padded, and align it in eccentricity 2 mm at the outer circle, ensure the eccentricity error within 0.01 mm; 3) Rough and finish turning inner hole to size $\phi 40_{+0.025}^{+0.050}$ mm and with a length of 15 mm; 4) Round off sharp edges of the inner hole	

No.	Operation step	Diagram
	Item 3	
Step 1	Extend the workpiece by 50 mm and clamp it after aligning	
	1) Turning end face; 2) Drilling the hole of $\phi22$ mm, and a length of 50 mm; 3) Turning the outer circle to size $\phi52$ mm, and a length of 140 mm; 4) Knurl $m0.3$, and a length of 40 mm; 5) Rough and fine turning the outer circle to size $\phi50_{-0.039}^{0}$ mm, and a length of 22 mm;	
	1) Rough and finish turning the inner hole to size $\phi30_{+0.025}^{+0.050}$ mm and with a length of 11 mm; 2) Rough and finish turning the inner hole to size $\phi36_{0}^{+0.057}$ mm and with a length of 4 mm; 3) Round off sharp edges of the inner hole and the outer circle; 4) Cut the workpiece off to guarantee a length of 40 mm	
Step 2	Pad the workpiece with copper sheets, and clamp the outer circle of size $\phi50$ mm	
	1) Turning the end face to ensure that the total length of the workpiece is 39 mm; 2) Rough and fine turning the inner hole to size $\phi24_{+0.025}^{+0.050}$ mm; 3) Turning the internal thread M27; 4) Rough and fine turning the outer circle to size $\phi50_{-0.039}^{0}$ mm and with a length of 7mm to ensure that the step length at the knurled part is 10 mm; 5) Chamfer $C1$ mm	

III. Workpiece quality test

1. Eccentric shaft

The eccentric shaft shall be tested according to its technical parameters shown in Figure 4.15.

1) Outer diameters, eccentricity and roughness

The outer diameters of $\phi30_{-0.028}^{-0.007}$ mm and $\phi40_{-0.050}^{-0.025}$ mm shall be tested according to the dimensional tolerance, and it is unqualified if out of tolerance.

The outer diameter is $\phi52$ mm, its the unmarked tolerance shall be tested according to GB/T1804—2000.

The eccentricity of 2 ± 0.01 mm shall be tested according to the dimensional tolerance, and it is unqualified if out of tolerance.

Roughness is $Ra1.6$ μm, or unqualified if downgrading.

2) Knurling

The knurling $m0.3$ texture should be clear and not chaotic.

3) Ordinary external thread

The $M27\times3$ thread should be tested with a thread ring gauge, and it is unqualified if out of tolerance.

4) Groove

The dimensions of the groove are 6 mm $\times2$ mm and $1\times\phi38$ mm, the unmarked tolerance should be tested according to GB/T 1804—2000.

5) Lengths

Lengths of the part are 10 mm, 14 mm, 31 mm and 85 mm, the unmarked tolerance shall be tested according to GB/T 1804—2000.

6) Others

The unmarked chamfer $C1$ mm is tested according to GB/T 1804—2000.

Other parts roughness is $Ra3.2$ μm, it is unqualified if out of tolerance.

2. Eccentric sleeve

The eccentric sleeve should be tested according to the processing requirements as shown in Figure 4.16.

1) Outer diameter and roughness

The outer diameters of $\phi 50^{0}_{-0.039}$ mm shall be tested according to the dimensional tolerance, and it is unqualified if out of tolerance.

Roughness is $Ra1.6$ μm, unqualified if downgrading.

2) Inner diameters, eccentricity and roughness

The inner diameters of $40^{+0.05}_{+0.025}$ mm and $\phi 30^{+0.025}_{0}$ mm and the inner diameters of grooves of $36^{0}_{-0.047}$ mm and $\phi 46^{+0.063}_{+0.021}$ mm should be tested according to the dimensional tolerance, and it is unqualified if out of tolerance.

The eccentricity of 2 ± 0.01 mm shall be tested according to the dimensional tolerance, and it is unqualified if out of tolerance.

Roughness is $Ra1.6$ μm, unqualified if downgrading.

3) Lengths

The lengths are 15 mm and 4.5 mm, their unmarked tolerance shall be detected according to GB/T 1804—2000.

The length $40^{+0.1}_{0}$ mm shall be tested according to the dimensional tolerance, and it is unqualified if out of tolerance.

4) Others

The chamfer is $C1$ mm, its unmarked tolerance shall be tested according to GB/T 1804—2000.

Other roughness is $Ra3.2$ μm, and it is unqualified if out of tolerance.

3. Thread sleeve

The thread sleeve shall be tested according to the processing requirements shown in Figure 4.17.

1) Outer diameters and roughness

The outer diameters of $\phi 50^{0}_{-0.039}$ mm and $\phi 46^{0}_{-0.033}$ mm should be tested according to the dimensional tolerance, and it is unqualified if out of tolerance.

Another outer diameter is $\phi 52$ mm, its unmarked tolerance shall be tested according to GB/T 1804—2000.

Roughness is $Ra1.6$ μm, and unqualified if downgrading.

2) Inner diameters, eccentricity and roughness

The inner diameters of $\phi 30^{+0.050}_{+0.025}$ mm and $\phi 36^{+0.057}_{0}$ mm should be tested according to the dimensional tolerance, and it is unqualified if out of tolerance.

Roughness is $Ra1.6$ μm, and unqualified if downgrading.

3) Knurling

The knurling $m0.3$ texture should be clear and not chaotic.

4) Inner thread

The M27 × 3 thread should be tested with a thread ring gauge, and it is unqualified if out of tolerance.

5) Lengths

The lengths are 10 mm, 39 mm, 7 mm, 7 mm and 4 mm, their unmarked tolerance shall all be tested according to the GB/T 1804—2000.

6) Others

The chamfer is $C1$ mm, its unmarked tolerance shall be tested according to GB/T 1804—2000.

Other roughness is $Ra3.2$ μm, and it is unqualified if out of tolerance.

4. Assembly

If the three items cannot be assembled together, it is considered as unqualified assembly.

Ⅳ. Notes

(1) When turning eccentric parts, pay attention to using appropriate clamping force to avoid damaging the parts.

(2) The actual eccentricity values of Item 1 and Item 2 should be consistent, otherwise, assembly difficulties or uneven appearance contours may occur.

Subtask 2 Machining Cone Eccentric Four-item Assembly

【Knowledge and Skills Objectives】

(1) Mastering the method of determining datum parts in complex fitting.

(2) Master the method of measuring inner and outer cone diameter dimensions.

(3) Master the method of processing the four-item combination (parts assembly).

【Related Knowledge】

I . Diagram of cone eccentric four-item assembly

1. Diagram of the cone eccentric four-item assembly

The diagram of the cone eccentric four-item assembly is shown in Figure 4.22. The cone eccentric four-item assembly consists of four parts(items). Their fit types include eccentric shaft hole fit, cone fit, unified fit of multiple outer circles and inner holes, length dimension clearance fit of four items, coaxiality of outer diameter dimension, etc. The processing work includes the machining of the datum part, the sequential machining of each item, and the machining after assembly. Turning the conical surface requires calculating and aligning the eccentricity.

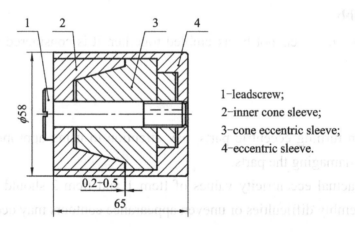

1-leadscrew;
2-inner cone sleeve;
3-cone eccentric sleeve;
4-eccentric sleeve.

Figure 4.22   The diagram of cone eccentric four-item assembly

Item 3 is composed of four parts: conical surface, cylindrical surface, eccentric cylindrical surface, and inner hole, and the structure of inner hole is the most complex one. The machining deviation of the conical surface and inner hole will affect the fit of Item 3 and Item 1 (lead screw), and the fit of Item 1 and Item 2 (inner cone sleeve), such as causing the lead screw to pass through smoothly and affecting the fit gap between the lead screw and the cone sleeve. The machining deviation of the eccentric cylindrical surface will affect the fit of Item 3 and Item 4 (eccentric sleeve), causing the fit to be stuck or even unable to be assembled. Item 1 is the datum part and should be machined first.

2. The lead screw of the cone eccentric four-item assembly

Item 1 (lead screw) is used to assemble and lock Item 2, Item 3 and Item 4, its diagram is as shown in Figure 4.23.

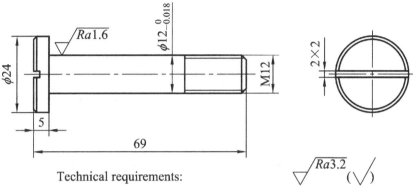

Technical requirements:

1. Unmarked Chamfering $C1$;

2. Unmarked tolerance is processed according to GB/T 1804−2000;

3. Round off sharp edges;

4. Allowed to connect.

Figure 4.23   The diagram of lead screw

3. The inner cone sleeve of the cone eccentric four-item assembly

The Item 2 (inner cone sleeve) is assembled with Item 1, Item 3 and Item 4, its diagram is shown as in Figure 4.24.

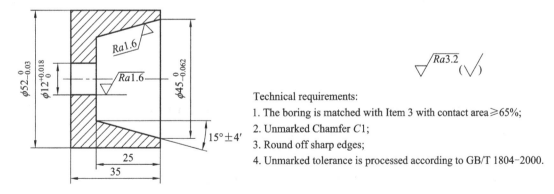

Technical requirements:

1. The boring is matched with Item 3 with contact area ≥65%;

2. Unmarked Chamfer $C1$;

3. Round off sharp edges;

4. Unmarked tolerance is processed according to GB/T 1804−2000.

Figure 4.24   The diagram of inner cone sleeve

4. The cone eccentric shaft of the cone eccentric four-item assembly

The Item 3 (cone eccentric shaft) is assembled with Item 1, Item 2 and Item 4, its diagram is shown as in Figure 4.25.

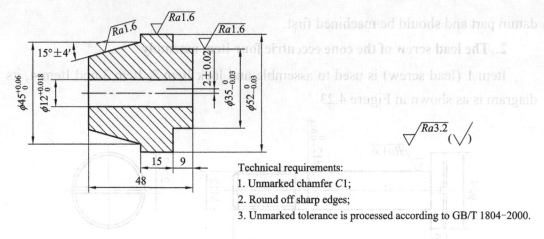

Technical requirements:
1. Unmarked chamfer $C1$;
2. Round off sharp edges;
3. Unmarked tolerance is processed according to GB/T 1804−2000.

Figure 4.25   The diagram of cone eccentric shaft

5. The eccentric sleeve of the cone eccentric four-item assembly

Item 4 (eccentric sleeve) is assembled with Item 1, Item 2 and Item 3, its diagram is shown as in Figure 4.26.

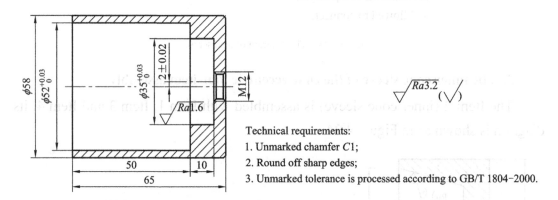

Technical requirements:
1. Unmarked chamfer $C1$;
2. Round off sharp edges;
3. Unmarked tolerance is processed according to GB/T 1804−2000.

Figure 4.26   The diagram of eccentric sleeve

II. The method of measuring the minor end diameter of the cone by two measuring rods with equal diameter

The major end diameter of the cone can be measured indirectly according to the minor end diameter of the cone. When the minor end diameter of the cone is measured, it can be determined whether the major end diameter of the cone is qualified. The diagram of measuring the minor end diameter of the cone with two measuring rods is shown in Figure 4.27.

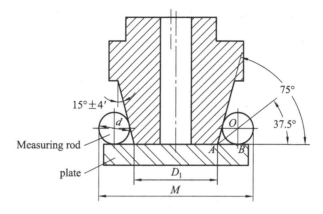

Figure 4.27   The diagram of measuring the minor end diameter

of the cone with two same measuring rods

Place the measuring rods symmetrically, and place the end face of the workpiece on the plate, measure the outer diameter of the measuring rod at both ends with a micrometer to obtain the M value.

The clamped angle of each of the measuring rod is $90° - 15° = 75°$, and the minor end diameter of the cone value is calculated by $\triangle ABO$.

As is known to all,

$$BO = \frac{d}{2}$$

$$AB = \frac{d/2}{\tan(75°/2)}$$

$$D_1 = M - d - 2AB = M - d - 2\frac{d/2}{\tan(75°/2)}$$

$$= M - d - d\frac{1}{\tan(75°/2)}$$

$$= M - d\left(1 + \frac{1}{\tan(75°/2)}\right)$$

$$= M - d\left(1 + \frac{1}{\tan 37.5°}\right)$$

In the above formula, D_1: the minor end diameter of the cone, in mm; M_1 : the reading value of the measuring rod, in mm; d: the diameter of the measuring rod, in mm.

Ⅲ. The simple and indirect method for measuring the major end diameter of the taper hole

The schematic diagram of measuring the major end diameter of the taper hole by measuring the edge size of the taper hole is shown in Figure 4.28.

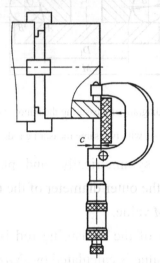

(a) Put the micrometer against the standard steel plate to measure the wall thickness value c

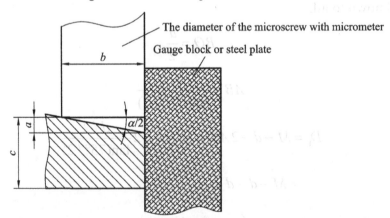

(b) Measure the diameter of the micro-screw of the micrometer to calculate the major end diameter of the taper hole

Figure 4.28　The schematic diagram of measuring the major end diameter of the taper hole

Use a standard steel plate against the end face of the workpiece, and measure the reading value with the micrometer inside against the steel plate, as shown in Figure 4.28(a). The reading principle is shown in Figure 4.28(b). The diameter of the micro-screw of the micrometer is b. Assuming that the cone angle is α, the oblique angle is $\alpha/2$, the short side $a = b \times \tan\dfrac{\alpha}{2}$, the outer diameter is d, and the major end diameter of the taper hole is D, there is

$$D = d - 2c + 2b\tan\frac{\alpha}{2} = d + 2\left(b\tan\frac{\alpha}{2} - c\right)$$

Ⅳ. Method of measuring the inner cone angle

The inner cone angle $15° \pm 4'$ can be tested with a sine gauge as follows:

(1) Calculate the height H of the gauge block group. The diagram of measuring the inner cone angle is shown in Figure 4.29, first rotate the sine gauge angle, insert the gauge block group.

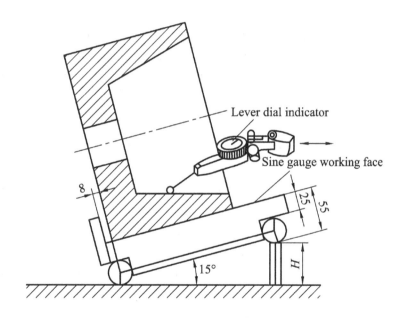

Figure 4.29   The diagram of measuring the inner cone angle

Example 4.1: Given that the half angle of the cone is $15°$ and the center distance L of the sine gauge is 200 mm, please calculate the height H of the gauge block that needs to be padded into.

Solution: Height of the inserted gauge block $H = L\sin\alpha = 200\sin15° = 200 \times 0.259 = 51.764$ mm.

(2) Place the parts on the sine gauge, insert the gauge blocks with a height of 51.76mm, press the lever dial indicator onto the surface of the inner cone, use the

plate as the datum plane, move the contacts inside and outside, and detect the angle of inclination.

【 Skills Training 】

Processing Cone Eccentric Four-item Assembly

Processing the cone eccentric four-item assembly as shown in Figure 4.22 to Figure 4.26.

Ⅰ. Operation preparation

The operational preparation for machining the cone eccentric four-item assembly is shown in Table 4.11.

Table 4.11   Operation preparation for machining the cone

eccentric four-item assembly

No.	Name		Preparation
1	Material		Medium carbon steel ϕ30 mm × 100 mm, ϕ55 mm× 100 mm, ϕ60 mm × 80 mm
2	Device		CA6140 lathe
3	Processing tools	Cutting tools	90° turning tool, 45° elbow turning tool, 90° inner hole turning tool, inner hole light tool, cutting tool, reticulate knurling cutter tool m0.3, center drill A2.5/6.3, tap die M12, drill bits of ϕ10 mm, ϕ10.3 mm, ϕ11.7 mm, ϕ27 mm, ϕ29 mm, and ϕ48 mm, reamer ϕ12H7 mm
4		Measuring tools	Feeler gauge, vernier caliper 0.02 mm / (0−150 mm), outside micrometer 0.01 mm / (0−25 mm, 25−50 mm, 50−75 mm), inside dial gauge 0.01 mm / (35−50 mm, 50−160 mm), universal angle gauge 2′/ (0° −320°), 60° pitch gauge, magnetic base dial gauge (0−6 mm)
5		Others	Red lead powder, marking disc, slotted screwdriver, adjustable wrench, tops and drilling fixture, other common tools

Ⅱ. Operation procedure

The operation steps for processing the cone eccentric four-item assembly are shown in Table 4.12.

Table 4.12   Operation steps for machining the cone eccentric
four-item assembly

No.	Steps	Diagram
	Item 1	
Step 1	Extend the part $\phi 30$ mm $\times$ 90 mm with 20 mm stretching out and clamp it	
	1) Turning the end face; 2) Drilling the center hole	
	Clamp the workpiece with one clamp and one top method with 70 mm stretching out: 3) Rough turning the outer circle $\phi 13$ mm to $\phi 12_{-0.018}^{0}$ mm and with a length of 69 mm; 4) Fine turning the inner circle $\phi 12_{-0.018}^{0}$ mm, and with a length of 69 mm; 5) Chamfer $C1$ mm; 6) Tap the M12 thread with the split die M12	
Step 2	U-turn and clamp the workpiece	
	1) Turning the outer diameter $\phi 24$ mm; 2) Turning the flat end face with a length of 64 mm; 3) Chamfer $C1$ mm	
	Item 2	
Step 1	Extend the part ϕ 55 mm $\times$ 100 mm with 50 mm stretching out and clamp it	
	1) Turning the flat end face; 2) Rough and fine turning the outer circle $\phi 53$ mm to $\phi 52_{-0.03}^{0}$ mm and with a depth of 36 mm; 3) Drill the hole $\phi 10$ mm and with a depth of 40 mm; 4) Drill the hole $\phi 29$ mm and with a depth of 25 mm;	

No.	Operation steps	Diagram
Step 1	5) Rough turning the major end diameter $\phi44$ mm of the taper hole and with a length of 24 mm; 6) Finish turning the end face; 7) Finish turning the taper hole for $\phi45_{-0.062}^{0}$ mm and with a length of 25 mm, and use Item 3 to fit the cone to ensure that the contact area of the cone surface is not less than 70%, and the fit gap is 0.2−0.5 mm; 8) Expand the hole $\phi11.7$ mm and ream it to $\phi12_{0}^{+0.02}$ mm; 9) Finish the outer circle $\phi52_{-0.03}^{0}$ mm and with a length of 36 mm; 10) Chamfer $C1$; 11) Cut the workpiece off to ensure a total length of 36 mm	
Step 2	U-turn and clamp the workpiece	
	1) Turning the flat end face to ensure a length of 35 mm; 2) Chamfer $C1$	
Item 3		
Step 1	Clamp one end of the part $\phi60$ mm $\times$ 80 mm with 55 mm stretching out	
	1) Turning the flat end face; 2) Rough turning the outer circle $\phi53$ mm to $\phi52_{-0.03}^{0}$ mm and with a length of 50 mm	
	3) Drill the hole $\phi10$ mm; 4) Rough turning the cone to the major end diameter of $\phi46$ mm; 5) Expand the hole $\phi11.7$ mm and ream it to $\phi12_{0}^{+0.02}$ mm; 6) Finish turning the cone to the major end diameter of $\phi45_{0}^{+0.06}$ mm and with a length of 24 mm; 7) Finish turning the outer diameter of $\phi52_{-0.03}^{0}$ mm and with a length of 50 mm	

No.	Operation steps	Diagram
Step 2	U-turn the workpiece and clamp the outer circle $\phi 52_{-0.03}^{0}$ mm	
	1) Use a dial indicator to correct the eccentricity value in the range of (2 ± 0.02) mm and the circular runout value of the left end face to zero; 2) Turning the right end face to ensure a total length of 48 mm; 3) Rough and finish turning the diameter of the eccentric circle to $\phi 35_{-0.03}^{0}$ mm ; 4) Chamfer C1mm	
	Item 4	
Step 1	Extend the part $\phi 60$ mm $\times$ 80 mm with 68 mm stretching out and clamp it	
	1) Turning the flat end face; 2) Drill the hole $\phi 10.3$ mm and with a depth of 70 mm; 3) Drill the bottom hole $\phi 28$ mm and with a depth of 59 mm; 4) Drill the hole of $\phi 48$ mm and with a depth of 50 mm at the bottom of the part; 5) Rough turning the outer circle from $\phi 59$ mm to $\phi 58$ mm; 6) Rough turning the inner hole from $\phi 51$ mm to $\phi 52_{0}^{+0.03}$ mm; 7) Rough turning the eccentric inner hole at the bottom to $\phi 30$ mm; 8) Finish turning the inner hole $\phi 52_{0}^{+0.03}$ mm and with a depth of 50 mm; 9) Finish turning the outer diameter of $\phi 58$ mm	

No.	Operation steps	Diagram
Step 2	Clamp the workpiece and align the center eccentricity	
	1) Use the dial indicator to find the eccentric value in the range of (2 ± 0.02) mm; 2) Turning the eccentric inner hole to $\phi 35^{+0.03}_{0}$ mm and with a depth of 10 mm; 3) Cut the workpiece off for a length of 66 mm	
Step 3	U-turn the workpiece and clamp the outer circle	
	1) Turning the flat end face to guarantee a length of 65 mm; 2) Tap the M12 internal thread; 3) Chamfer $C1$	

III. Workpiece quality test

1. Lead screw

The lead screw shall be tested according to its technical parameters shown as in Figure 4.23.

1) Outer diameter and roughness

The outer diameter $\phi 12^{0}_{-0.018}$ mm shall be tested according to the dimensional tolerance, and it is unqualified if out of tolerance.

The outer diameter is $\phi 24$ mm, its unmarked tolerance shall be tested according to GB/T1804—2000.

Roughness is $Ra1.6$ μm, and unqualified if downgrading.

2) Length and groove

The length dimensions are 5 mm and 69 mm, their unmarked tolerance shall be tested according to GB/T1804—2000.

The groove dimension is 2 mm × 2 mm, its unmarked tolerance shall be tested according to GB/T1804—2000.

3) M12 thread

M12 thread is threaded with the split die M12, it is unqualified if the tooth shape is incomplete.

2. Inner cone sleeve

Test the inner cone sleeve according to its technical parameters shown in Figure 4.24.

1) Outer diameter and roughness

The outer diameter $\phi 52_{-0.03}^{0}$ mm shall be tested according to the dimensional tolerance, and it is unqualified if out of tolerance.

Roughness is $Ra3.2$ μm, and unqualified if downgrading.

2) Inner diameter, inner cone hole, and roughness

The inner diameter $\phi 12_{0}^{+0.018}$ mm and inner cone hole $\phi 45_{-0.062}^{0}$ mm shall be tested according to dimensional tolerance, and it is unqualified if out of tolerance.

Roughness is $Ra1.6$ μm, it is unqualified if downgrading.

3) Cone

The cone angle of $15° \pm 4'$ shall be tested according to the angle tolerance, and it is unqualified if out of tolerance.

4) Length

The length dimensions are 35 mm and 25 mm, their unmarked tolerance shall be tested according to GB/T1804—2000.

5) Other

Other roughness is $Ra3.2$ μm, it is unqualified if downgrading.

3. Cone eccentric shaft

Test the cone eccentric shaft according to its technical parameters shown in Figure 4.25.

1) Outer diameters and roughness

The outer diameters $\phi52^{+0.03}_{0}$ mm and $\phi35^{+0.03}_{0}$ mm shall be tested according to dimensional tolerance, and unqualified if out of tolerance.

Roughness is $Ra1.6$ μm, it is unqualified if downgrading.

2) Inner diameter and roughness

The inner diameter $\phi12^{+0.018}_{0}$ mm shall be tested according to dimensional tolerance, and it is unqualified if out of tolerance.

Roughness is $Ra1.6$ μm, and unqualified if downgrading.

3) Cone

Major diameter of outer cone $\phi45^{+0.06}_{0}$ mm shall be tested according to dimensional tolerance, and it is unqualified if out of tolerance.

The cone angle of $15° \pm 4'$ shall be tested according to the angle tolerance, and it is unqualified if out of tolerance.

4) Length

The dimensions of length are 48 mm, 15 mm, and 9 mm, their unmarked tolerance shall be tested according to GB/T 1804—2000.

5) Other

The remaining chamfers are $C1$, its unmarked tolerance shall be tested according to GB/T 1804—2000.

Other roughness is $Ra3.2$ μm, and it is unqualified if downgrading.

4. Eccentric sleeve

Test the eccentric sleeve according to its technical parameters shown in Figure 4.26.

1) Outer diameter, inner diameters, and roughness

The outer diameter is $\phi58$ mm, its unmarked tolerance shall be tested GB/T 1804—2000.

The inner diameters $\phi52^{+0.03}_{0}$ mm and $\phi35^{+0.03}_{0}$ mm shall all be tested according to dimensional tolerance, and unqualified if out of tolerance.

Roughness is $Ra1.6$ μm, it is unqualified if downgrading.

2) Eccentricity

The eccentricity of 2 ± 0.02 mm shall be tested according to the dimensional tolerance, and it is unqualified if out of tolerance.

3) Internal thread

M12 internal thread shall be be threaded with the split die, and incomplete thread profile is not qualified.

4) Length

The length dimensions are 50 mm, 65 mm, and 10 mm, their unmarked tolerance shall be tested according to GB/T1804—2000.

5) Other

The remaining chamfers are $C1$, its unmarked tolerance shall be tested according to GB/T1804—2000.

Other roughness is $Ra3.2$ μm, it is unqualified if downgrading.

References

[1] XU Z F, LIANG J H. Lathe Technology [M]. 4th ed. Beijing: China Machine Press, 2004.

[2] XIAO Y C, HAN Y S. Turner (Elementary)[M]. 2rd ed. Beijing: China Labor and Social Security Press, 2013.

[3] SUN Y L. Mechanical Cutting Skills [M]. Beijing: China Machine Press, 2000.

[4] ZHAO F, WANG Z Y, FANG L. Machinery Manufacturing Foundation [M]. Tianjin: Tianjin University Press, 2012.